Deep Study of the Universe through Torsion

Deep Study of the Universe through Torsion

By

Francisco Bulnes

Cambridge Scholars Publishing

Deep Study of the Universe through Torsion

By Francisco Bulnes

This book first published 2022

Cambridge Scholars Publishing

Lady Stephenson Library, Newcastle upon Tyne, NE6 2PA, UK

British Library Cataloguing in Publication Data
A catalogue record for this book is available from the British Library

ISBN (10): 1-5275-8728-2
ISBN (13): 978-1-5275-8728-1

CONTENTS

Appendices:

ACKNOWLEDGEMENTS

The Author would like to thank Helen Edwards of CSP for her attention and professionalism in the difficult process of editing and publication; her enthusiasm and capability has helped establish the scheduling and pertinent strategies in the publication of this work.

ACKNOWLEDGEMENTS

The Author would like to thank Helen Edwards of CSP for her attention and professionalism in the difficult process of editing and publication; her enthusiasm and capability has helped establish the scheduling and pertinent strategies in the publication of this work.

PREFACE

A torsion is the observable result of a field interacting in space-time resulting in the creation of gravity from a vacuum-energy interaction. This leads to agitation in space-time resulting in leptogenesis and the creation of dark matter and dark energy, relevant to the process of baryogenesis and the evolution of the actual universe. Factors concerning inflation and matter-energy are analyzed, as are hadrons, which are fundamental to dark matter production. Some results of torsion with baryogenesis are given to analyze the evidence relating to baryon acoustic oscillations, the role of sterile neutrinos, and hadron proliferation in this step of the universe's development, as well as the different effects of cosmic expansion, redshifts, and space distortions etc., which are effects of gravity caused by torsion and matter production in the universe. Two results are obtained, one in the context of Majorana neutrinos and another that goes beyond neutrinos and deals with dark energy. The present monograph on torsion as a field observable has the fundamental goal of creating a specialized study of torsion and describes a new theory based on the energy of the curvature of the universe. Its applications are corollaries of this theory.

Prof. Dr. Francisco Bulnes, Author
IINAMEI Director
Head of Research Department in Mathematics and Engineering, TESCHA

CHAPTER 1

INTRODUCTION

In the study of the origin of our universe, the currently accepted model is based on the CMB[1]. In this model, we consider nucleosynthesis to be the fundamental process forming energy-matter. This is a long process taking several thousands of million years to pass through the various stages of leptogenesis and baryogenesis in the early universe and up to the proliferation of hadrons and tachyons, among other particles. Through the field mechanisms and interactions between these particles, certain oscillations are achieved at the fundamental level.

Over a period of time, the universe was formed with a large quantity of leptons of various classes of fermion, which developed the charges of the universe. Their interactions with gluons and other gauge bosons formed energy-matter through the torsion field[2] (with one of the products being gravitational waves), later passing this energy on to matter. A torsion, as an invariant of energy-space, is defined as $T^{\gamma}_{\alpha\beta}=2S^{\gamma}_{\alpha\beta}$ and involves energy through waves called spinors; it operates on points in space-time defined by the spinor correspondence [1] (see fig. 1). This process gives the appearance of waves agitating space-time as the field acts, at the microscopic scale, leading to the existence of energy in space. This energy can be an indication of gravity, measured by scientists as gravitational waves and involving other aspects of the field equations [2, 3].

---

[1] Cosmic microwave background.

[2] In this process of matter, the gravitational field is produced due to the actuation of the torsion field [1-3].

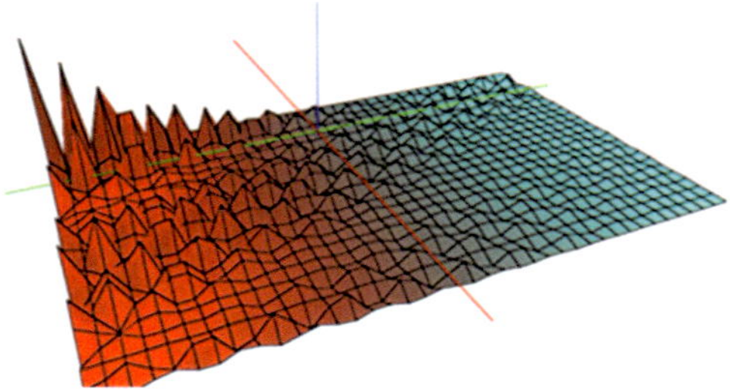

Fig. 1. Spinor image of spin-waves in torsion: 2-D model of spin-waves generated for a magnetic dilaton with a cylindrical-spiral movement trajectory [3-5].

The torsion field mechanism is fundamental to all the evolutionary processes of sidereal objects in the universe, such as galaxies, black holes, supernovas, pulsars, and nebulae, which are formed by the intersidereal field. For example, in galaxies the spiral arms are oriented and the galaxy's rotation is driven by the intersidereal magnetic field. However, the torsion field also results in the production of large amounts of positive energy originating at the microscopic level in early matter (atoms and primal particles, i.e. the so-called ***H***-particles[3]), which produce the mechanism of movement from the space-time vacuum and the energy interaction [6] (fig. 2).

[3] Higgs Bosons.

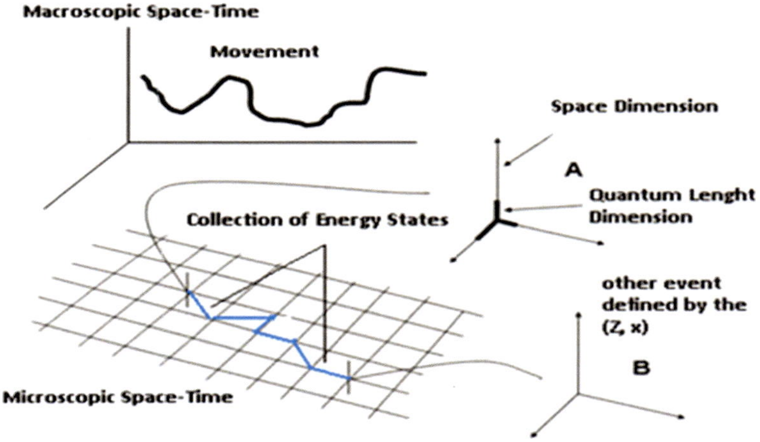

Fig. 2. Description of the ***H***-particle states.

In astrophysics, we study the accretion and rotation lines in galaxies, the movement of black holes, and their action in the universe. Likewise, the universe in its early years appears very similar to the actual universe with a high proliferation of atoms with extra energy, as evidenced for neutrinos (which constitute the major part of the universe) and free photons observed, for example, in the luminous disk of a galactic horizon up to 10 kpc (see fig. 3). However, this indicator is not precise according to the inhomogeneity detected by WMAP (redshift-space distortions and microwaves in background) observations of the background of the actual universe [4]. Here we may precisely consider an extension of the Standard Model; however, what happens with the gravitational failure in these effects of cosmic inhomogeneity [5]? In this respect, it is necessary to consider additional studies on the evidence for torsion and extrapolate this to the evolution of baryogenesis so as to prove that the effects of cosmic inhomogeneity are due to the action of the torsion field at the microscopic level of observation [6], for example, in terms of baryon oscillations within cosmic expansion (see fig. 3).

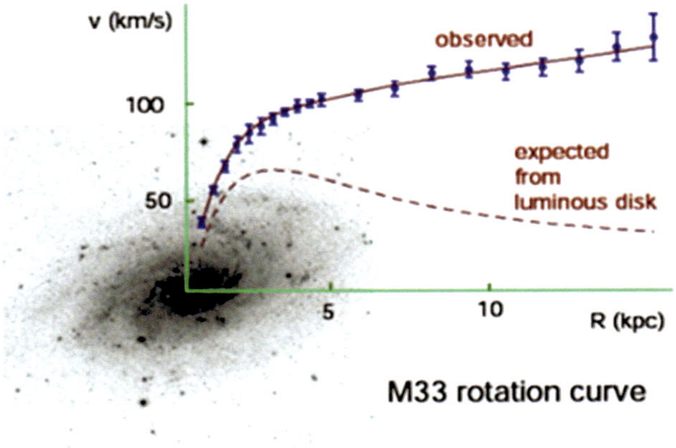

Fig. 3. Cosmic expansion in dark matter.

Some aspects of general relativity and quantum field theory will possibly be very useful and can be modified to fit with the CMB-model and the supersymmetrical extensions that must be considered in the Standard Model for field actions [7, 8]. Some of this model goes beyond the theory of neutrinos, for example, in the energy-matter context.

Cosmic expansion is equivalent to baryon oscillations, which produce dark energy. This fact can be gauged by considering an appropriate *w*-parameter inside the field equations in the framework of scalar field theory and accounting for certain values of *w* in relation to different physical characteristics or different key values[4]. For example, the values of $w = 0$ for matter; $w = 1/3$ for radiation; $w = -1$ for the cosmological constant; and $w < -1/3$ for the acceleration of fluids in the universe.

---

[4] Using baryon acoustic oscillations, it is possible to investigate the effect of dark energy in the history of the universe and constrain the parameters of the equation of state of dark energy.

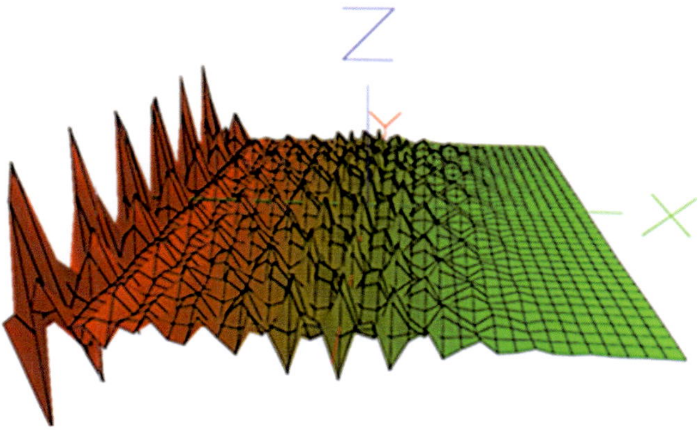

Fig. 4. Geometrical model created to explain the persistent nature of the torsion field in the production of gravitational waves [9]. Dark energy may have been produced from the remains of axions in the expansion process and is possibly a key value arising in the field equations. We can observe here the horizon of the 2-D space model and compare it to the 6-degree field galaxy redshift survey given in figure 3. An inconsistency may arise from the second process, or baryogenesis, through which dark energy joined with gravity at the beginning of the universe. This explains the appearance of matter and the start of the process of nucleogenesis [1].

The precursor values are considered characteristic of the corresponding work that is required for each physical entity (matter, radiation,.....), considering the latent force of certain "pressures" (tug) of the physical entity contrary to the displacement of certain heavy particles. These values are theoretical and are obtained through dynamic models in which the virial energy model of the universe is applied. It thus could be the case that dark energy has negative pressure, in contrast to dark matter, which has positive pressure.

Table I: Physics of Dark Energy with *w*-parameter and Scalar Factor *a*

#	Physics	Equation
1	State (radiation, temperature, scattering)	$P = w\rho$
2	Conservation of Energy	$dE = d(\rho a^3) = -pd(a^3)$
3	Rearrangement	$\rho \propto a^{-3(1+w)}$
4	Friedmann equation (thermal inflation, expansion, matter variation)	$da/dt \propto a^{-(1+3w)/2}$

Finally, the variations in pressure in both physical aspects—dark matter and energy—can be detected throughout the universe's expansion process as baryon oscillations, which has been shown by experiments with heavy particles. As such, *dark energy is gravitational + photon energy*, filling space and interacting with matter-producing field effects from these bodies, which are observed as maser and laser radiation. In fact, in this step of the universe there is a proliferation of **H**-particles[5], which can explain the large quantity of hydrogen in the universe, it being a fundamental chemical element composing interstellar and sidereal bodies (stars, globular clusters, nebulae, galaxies…).

[5] Hydrogenic atoms.

CHAPTER 2

OBSERVATIONS, EXPERIMENTS, AND MEASUREMENTS OF TORSION AS A FIELD OBSERVABLE

2.1. Geometrical Models and Signal Analysis

A torsion is a double curvature of a space or body resulting from the interaction of two fields—one field is the electromagnetic field and the other is a field that is relative to matter, i.e. the gravitational field in its broad sense (even considering quantum gravity).

Likewise, as mentioned in the introduction, a torsion born from the spins of matter and under the action of the electromagnetic field, produces spin-waves whose geometrical invariants are spinors in the invariant theory [1, 4] (see fig. 5).

This has been formulated in an initial conjecture.

Conjecture 2.1.1. The curvature in the spinor-twistor framework can be perceived with the appearance of a torsion and anti-self-dual fields [1].

As stated in an electronics study [3] using a magnetic Hall effect sensor[6], evidence of the existence of torsion as a field observable was obtained using a magnetic particle as a dilaton [4, 11] moving through a trajectory whose torsion is constant in all space [12]. Likewise, the signal below was

[6] **Lemma [10].** We consider a Hall effect sensor device $\mathcal{L}^{H}_{\text{Hall}}$. The current deflection detected for the change in the magnetic field by the sensor produces torsion energy per unit of volume

$$\tau = \frac{V}{2\pi}\frac{b}{(a^2+b^2)}\frac{1}{1}\left(=\frac{\text{Volts}}{(\text{meter})^3}\right).$$

obtained[7]

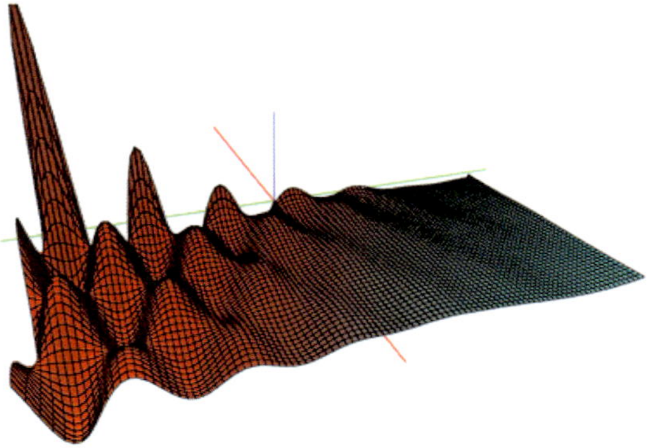

Fig. 5. Spinor image of spin-waves in torsion. This is a 2-D model of spin-waves generated for a magnetic dilaton in a cylindrical-spiral movement trajectory [5, 10].

We can consider this 2-D surface $s(\omega_1, \omega_2)$ as a mesh of 100 points of density and the corresponding torsion model is obtained as the cylindrical solution *Z = exp(0.5x)BesselY(0.1x, 2y)sin(4x)+BesselJ(0.6x, 6)* of the corresponding field equation $(\square + \chi T)\phi = 0$.

$$\tau(\omega) = \frac{1}{2\pi}\frac{b}{l^3}\int_{-0.5\,s}^{+0.5\,s}\Pi\left(\frac{t}{l}\right)e^{-j\omega t}dt = \frac{1}{2\pi}\left[\frac{(2.5\,V - \tilde{\alpha})b}{l^3}\right]\frac{\sin(0.5\ \omega)}{0.5\,\omega} \qquad (1)$$

[7] A torsion is detected with conditions of movement. Likewise, in line with lemma 3.1 [3], the magnetic field produced in the dilaton must correspond to

$$H = \frac{I}{2\pi}\frac{a^2}{l^3}$$

and the magnetic field of the dilaton must decrease in cycles of a few seconds. These are detectable using the Hall effect sensor (with low velocity). Thus, the initial constant voltage signal is conditioned as

$$V_0 = \Pi\left(\frac{t}{13}\right) = \begin{cases} 2.5V - \alpha & |t| \leq 0.5\,s \\ 0 & |t| \geq 0.5\,s \end{cases}$$

presenting a rectangular signal. Here, $\tilde{\alpha}$ is an amplifying factor of the voltage so that it is detectable.

which belongs to the set of signals that show torsion under permanent torsion conditions

$$\left\{\frac{\sin\omega L}{\omega L},\frac{\sinh\omega L}{\omega L};\frac{\cos\omega L}{\omega L},\frac{\cosh\omega L}{\omega L}\right\},\forall L=\frac{n2\pi}{\omega T},T>0 \quad (2)$$

In our experiments, we have only very limited detection of spin/waves and we need to use signal filters to obtain a signal in a time range of at least 10 seconds. The corresponding screen and 2-dimensional spinor surface $s(\omega_1,\omega_2)$ can be viewed in figures 6a, 6b, and 7b. The oscillations $s(\omega_1,\omega_2)$ are small and are in two planes.

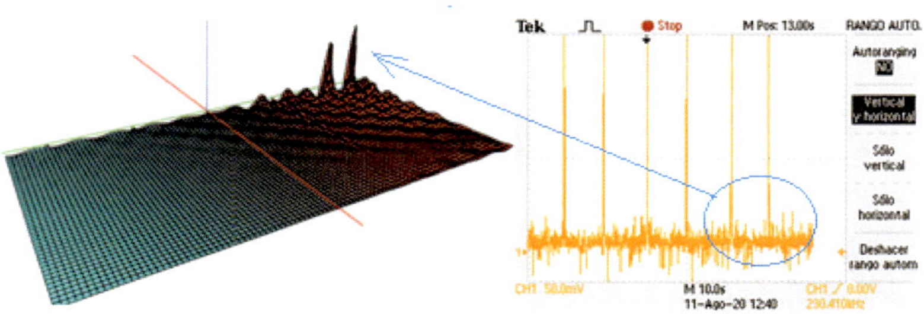

A) B)

Fig. 6. A) The corresponding solution to the field equation in the cylindrical regime is Z = exp(-0.5(x+1))BesselY(0.5x,8y)sin(7x)BesselJ(0.3y,20). The rotation of the cylinder was realized as in B). B) Corresponding signals in torsion detection and shown by the wave generator under the following measuring conditions: frequency of 235 kHz; wave sample of 13 seconds; and voltage of 4 V (fig. 7 A)).

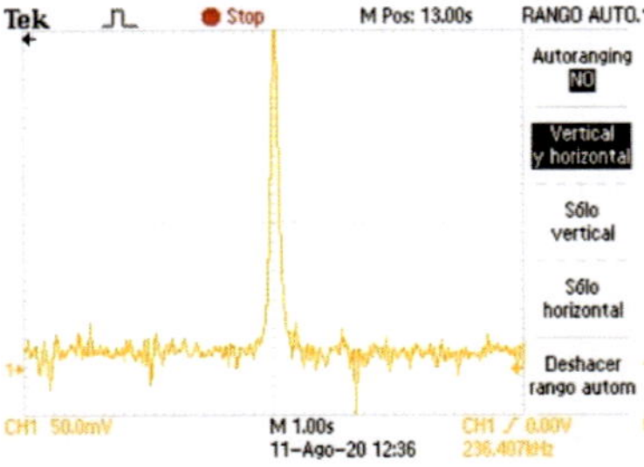

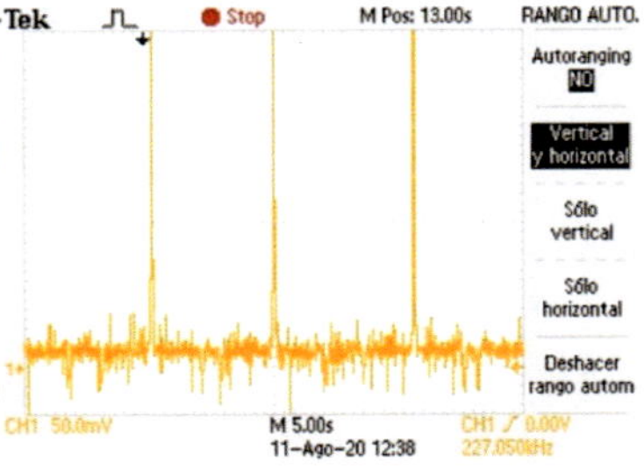

A) B)

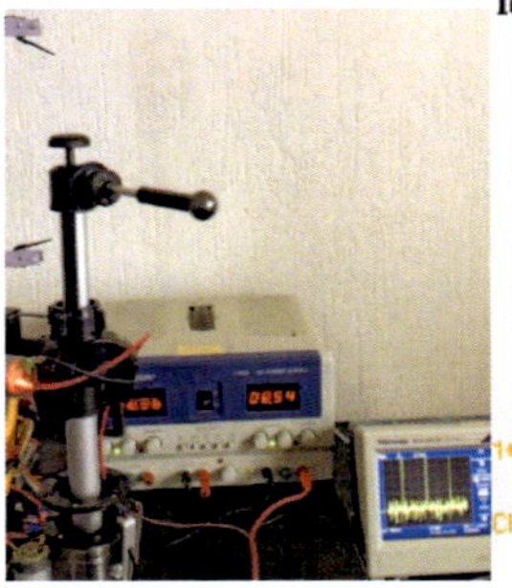

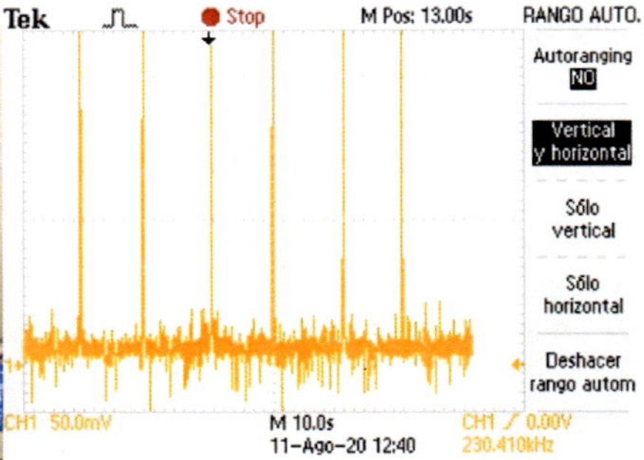

C) D)

Fig. 7. A) Energy spectra obtained for the conditioning of the voltage signals. B) This represents a quasi-periodic signal where for each 14 *seconds* we see a pico-voltage V_p, when the dilaton spins to a velocity of 12 *rpm*, we can observe that the amplitudes are greater than the subsequent amplitudes until we get a new maximum V_p. C) The frequency ω is maintained in the range 227 to 236 kHz. In the sensing process, energy spectra of 1*sec/div* (A), closer to *5sec/div* (C, D), and 10*sc/div* (D) to 50 *mV* for the three cases were realized. The signal was conditioned through the use of the AD620 instrumentation amplifier, which permits us to view an output voltage proportional to 1.4 *mV* per *Gauss* detected.

The dilaton or particle reveals a torsion in space-time [4, 13] under the action of an electromagnetic field and considering the kinematic invariance. Then, we choose a trajectory of constant torsion for our electronic device (figs. 8 & 7 D)) [13].

Conjecture 2.1.2 (F. Bulnes) [3, 5]. Torsion is the geometrical invariant of the interaction between energy and space.

The appearance of waves in space-time agitation is due to the energetic agitation of particles that, through their spins, interchange through the screw effect generated by the force $\delta(t_1, t_2)$ (see fig. 8).

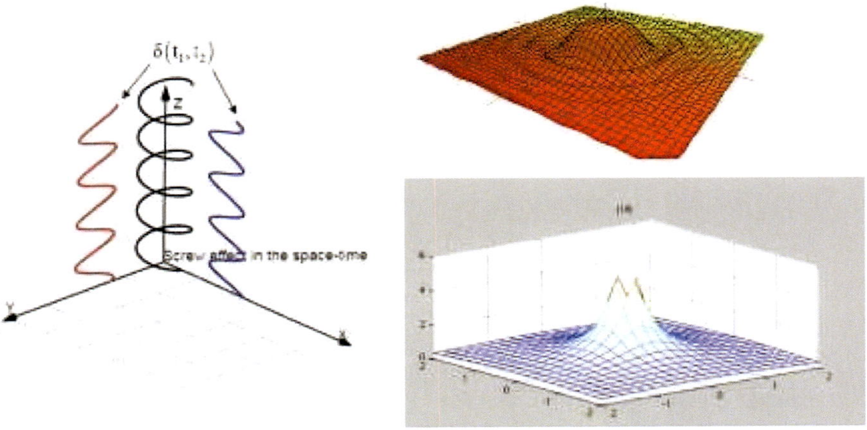

A) B)

C)

Fig. 8. A) The screw effect in space-time. B) The monopole of the electromagnetic field. This represents only $\delta(t_1, t_2)$ of a field source in space-time, whose torsion is established on the Z-axis. C) The 2-D surface shows the field source with the effect of torsion creating waves in the 2-D surface around the source.

In studies of the universe, the creation of geometrical models or numerical simulations of space-time phenomena is extremely complicated and requires reinterpretation to obey modern field theories and the cohomological theory of topological spaces, including the incorporation of microscopic theories of the universe as this is the level at which we find the root causes of all these phenomena. However, as has been stated, the most important geometrical invariant is curvature and torsion is the "second curvature". For example, we present the following reinterpretation.

The screw-effect geometrical model and a reinterpretation of cosmological objects such as black holes, sources such as stars, or the intersidereal magnetic alignment of galaxies using 2-dimensional complex surfaces and the Morera and Cauchy-Goursat theorems to evaluate them, can applied through a numerical program. Likewise, for example, in relation to singularities or poles in space-time, we can consider space-time a complex Riemannian manifold with singularities. This can represent the surface of the real part of the function $g(z) = \frac{z^2}{z-1}$. The modulus of this point is at least 1 and thus lies inside one contour. Likewise, the contour integral can be split into two smaller integrals using the Cauchy-Goursat theorem, finally having the contour integral [12] $\oint_C g(z)dz = \oint_C \left(0 - \frac{1}{z - z_1}\right)dz = 0 - 2\pi i = -2\pi i$ (see fig. 9). Additionally, this value is a traditional cohomological functional element of $H^f(\Pi - \ell', \Omega^r) = \mathbb{C}$. This element is a contour around the singularity, as can be seen in figure 9.

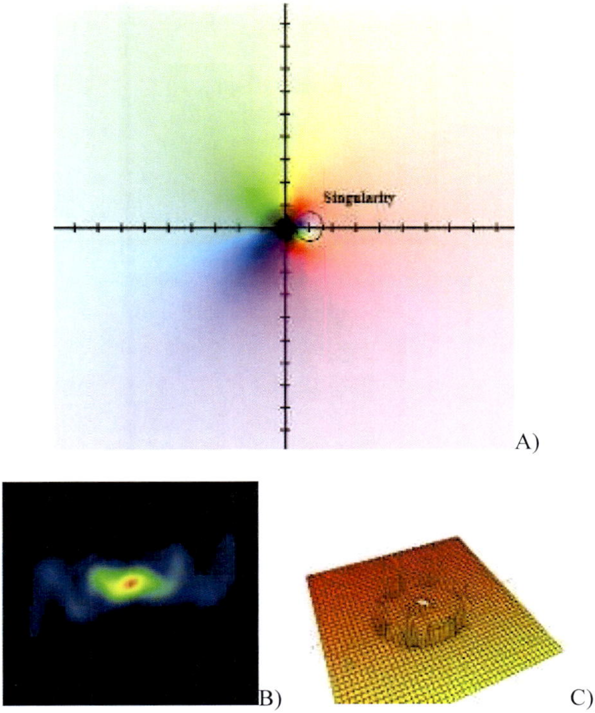

Fig. 9. A) Pole or singularity of the complex function $g(z) = \dfrac{z^2}{z-1}$. B) The black hole with astronomic catalogue number SS 433 is a giant black hole that wobbles. Using the VLA (Very Large Array) and the VLBA (Very Large Baseline Array), we can observe the wiggle of its jets over time. In the spectral image, we can see the corkscrew appearance obtained for the screw effect. This appearance is due to the agitation produced by the gravitational and electromagnetic fields that result from particles surrounding (right-handed neutrinos, keV) the black hole in vacuum of space. C) A 2-dimensional model considering small perturbations in the horizon detected in neighborhood of the source (this can also be considered as a type of singularity, for example, a peak or cusp). The surface is Z = *(0.2coth (-0.2ln (0.3x^2+0.2y^2)))/(x^2+y^2)*, where the perturbation phenomena occur. This surface represents the quantum field fluctuations of the beginning of the universe from the Big Bang up to any stage of space-time evolution before baryogenesis.

In cosmological evolution, torsion plays a fundamental role in the forming of all interstellar objects and even the universe itself. The macroscopic density fluctuations are echoes of the Big Bang right up to the large CMB we find today (quantum field fluctuations shown in fig. 9).

From a purely cosmological point ofview, the existence of these singular points reveals the existence of sidereal objects where large flows (using Poincarè arguments) of energy are expelled and/or attracted along space-time as twistors satisfying the geometrical Penrose model of a black hole. Thus, the fundamental twistor equation is satisfied. Torsion, as a field observable, is always present. As such, the spins s and $-s$ correspond to different field interactions whose images can correspond to the twistor spaces $\mathbb{P}^+$ and $\mathbb{P}^-$.

Here, we can incorporate the following space cohomology, which reinterprets field theory objects as geometrical objects and consider the universe as a complex Riemannian manifold (topological space) in relation to a field source

$$H^1(\mathbb{P}T, O(-2h-2)) \cong \ker(U, \square_{h(k)}) = \{\phi \in C^2(U) \mid \square_{h(k)}\phi = 0, \text{in}\, U \subset \mathbb{M}\}$$

The results of field torsion are evident in the twistor-spinor framework [1, 5] and we can use electronic interpretations from other experiments that extend the design of the electronic experiments realized in our research.

2.2. Electronics Experiments and Tensors Related to Torsion

The fundamental problem considered in this chapter is linked to the determination of energy-(space-)time variations that occur in the interaction of movement and matter-energy in a special geometry of movement or movement kinematics. However, we need a background component that permits measurement and detection under the invariance of its fields with the change in the spin of particles of matter, as can be seen in the case of torsion [10] considering a quasi-local matter model represented through gravitational waves of cylindrical type for the measurement and detection of the field torsion. In considering only a single component of the geometrical torsion that does not vanish along the curve of a particle, this last aspect gives us an object of study that moves and is affected by the radiation of energy, thus permitting the use of some physical effect such as the Hall effect.

The gauging of the torsion system through the deformation of space and using movement in an external field acting on a particle in motion offers a clear example of the dual concepts of the twistor frame and the spinors. The objective is to demonstrate the existence of the kinematic twistor tensor in a system that detects torsion. Then, we can obtain an image of the spinors due to the duality demonstrated in [1].

We know that we need an intermediate gauge field to experimentally establish the relation determined in [14] between the kinematic twistor tensor and the energy-matter tensor (this last is due to movement in space-time) in duality.

Likewise, we consider $\mathbb{M}$ space-time as a complex Minkowski model and we define the kinematic twistor tensor as that obtained in a region of space Σ considering the energy-matter tensor and whose image, as a 2-dimensional surface, will be the twistor 2-surface $\mathbb{T}(S)$. The geometrical evidence for torsion is given by this contorted surface.

Additionally, starting with the energy-matter tensor $T_{\alpha\beta}$, the kinematic twistor tensor $A_{\alpha\beta}$, in a radiation energy bath (electromagnetic radiation) will be defined by the interaction of two fields Z_1^{α} and Z_2^{α} that act in Σ

$$A_{\alpha\beta} Z_1^{\alpha} Z_2^{\beta} = \int_{\Sigma} T_{\alpha\beta} k^{\alpha} d\sigma^{\beta} \qquad (3)$$

which produces a total electrical charge due to the Gauss divergence theorem on the currents $T_{\alpha\beta}\, k^{\alpha}$

$$\mathrm{Q}[k] = \frac{1}{4\pi G} \int_{\Sigma} R_{\alpha\beta\gamma\delta} f^{\alpha\beta} d\sigma^{\gamma\delta} \qquad (4)$$

This can be identified as the source, depending on the Killing vector k^{α} of the Minkowski space background, modeled as

$$\mathbb{M} = S^2 \otimes \mathbb{C}^2 \otimes M \qquad (5)$$

where M is the space-time of two components

$$M = S^+ \oplus S^- \qquad (6)$$

As such, its system has a family of complex 4-dimensional solutions $(\cong \mathbb{C}^2)$ and the family defines the 2-surface twistor space $\mathbb{T}(S)$.

We can define the space of the kinematic twistor tensor as the space of tensors [1]

$$(\mathbb{T}(S) \odot \mathbb{T}(S))^* = \{A_{\alpha\beta} \in \mathrm{T}_2^4(M) | \mathrm{A}_{\alpha\beta} Z^{\alpha} Z^{\beta} = \mathrm{Q}[k]\} \qquad (7)$$

With a gauge field (electromagnetic field as photons) acting on the background radiation of the Minkowski space $\mathbb{M}$ and the energy of the matter, as has been said, will be related to this gauge field through the equation

$$j^{\alpha} = T_{\alpha\beta} k^{\alpha} \qquad (8)$$

where k^{α} represents the density of background radiation, establishing the curved part of space (with spherical symmetry) together with $T_{\alpha\beta}$ (see fig. 10).

$$\mathrm{Q}[k] = \frac{1}{4\pi G} \int_{S^2} T_{\alpha\beta} k^{\alpha} d\sigma^{\beta} \geq \int_{S^2} j^{\alpha} d\sigma^{\beta} \geq 2\pi\chi \qquad (9)$$

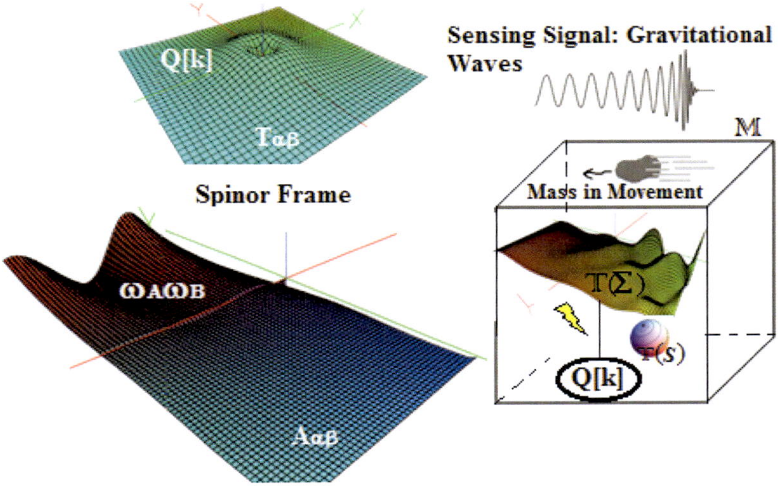

Fig. 10. *Supermassive mass movement field + electromagnetic energy field = evidence of torsion on the surface of the sensor (sphere).* How then can we construct a tensor that provides evidence of torsion using trace electronic signals that may be the result of gravitational waves from matter and electromagnetic fields? We need a tensor of the invariants of movement identified by invariants in geometry. This is the kinematic twistor tensor $A_{\alpha\beta}$. Space S is the sphere used to sense torsion and that transmits its variation at the specified time to the surface Σ defined by the movement of electromagnetic-matter. A 2-D model of the spinor representation of the kinematic twistor tensor $A_{\alpha\beta}$ can be constructed from the sphere.

The corresponding electromagnetic device generates an electromagnetic radiation bath in space and the movement of a mass inside this region produces variations in the electromagnetic field. Using a curvature energy sensor [14-16], we can obtain a spectrum in the twistor-spinor frame.

Likewise, according to twistor-spinor theory, and using the duality between the tensors $T_{\alpha\beta}$ and $A_{\alpha\beta}$, we can determine the mechanism of measurement and characterize the geometrical context of the detection of torsion. We define the twistor space by the points[8]

$$\mathbb{T} = \{Z^{\alpha} | Z^{\alpha} = (\omega^{A}, \pi_{A'})\} \qquad (10)$$

for all coordinate systems A and A'. We define the twistor infinity tensor $I_{\alpha\beta}$ [9] as that obtained directly for the whole of space-time whose structure obeys the rules of Minkowski space $\mathbb{M}$ and the surface Σ, which is a 3-surface, is the result of the twistor fields Z and Z^{β}. This is to say

$$\Sigma = \Sigma(Z^{\alpha}, Z^{\beta}) \qquad (11)$$

having a metric that is defined when $\alpha = \beta$ and $Z^{\beta} = \overline{Z^{\beta}}$ (its complex conjugate). Thus, in the infinity of space-time we have the following sequence of mappings

$$\mathbb{T} \xrightarrow{I^{\alpha\beta}} \mathbb{T}(S) \xrightarrow{I^{\alpha\beta}_{\Sigma}} \mathbb{T}(\Sigma) \qquad (12)$$

whose correspondence rule is

$$Z^{\alpha} \mapsto I^{\alpha\beta} S_{\beta\beta'} \overline{Z^{\beta'}} \mapsto I^{\alpha\beta} \Sigma_{\beta\beta'\beta} \overline{Z^{\beta'}} \qquad (13)$$

We consider the symmetrical part of the fields Z and Z^{β} given by the spinors ω^{AB}, which satisfy the valence-2 twistor equation

$$\nabla^{A}_{A'} \omega^{AB} = -i \in^{A(B} k^{C)}_{A'} \qquad (14)$$

---

8 $\omega^{A} : \mathbb{T}^{*} \rightarrow \mathbb{T}$ with the rule of correspondence of points of space-time $\pi_{A'} \mapsto ix^{AA'} \pi_{A'}$. Furthermore, we have its dual $\pi_{A'} : \mathbb{T} \rightarrow \mathbb{T}^{*}$, with the correspondence rule of points of space-time $\omega^{A} \mapsto -ix^{AA} \omega^{A}$. Likewise, the corresponding twistor spaces in this case are
$\mathbb{T} = \{Z^{\alpha} = (\omega^{A}, \pi_{A'}) | \omega^{A} = ix^{AA'} \pi_{A'}\}$ $\quad \mathbb{T}^{*} = \{W_{\alpha} = (\pi_{A}, \omega^{A'}) | \omega^{A'} = -ix^{AA'} \pi_{A}\}$

9 $I_{\alpha\beta} : \mathbb{T}^{*} \rightarrow \mathbb{T}$, with the correspondence rule $W_{\alpha} \mapsto Z^{\alpha} I^{\alpha\beta} W_{\beta}$.

which has a solution in 10-D space. We need to bind the space-region of our study to spinor-waves in 4-D space, that is to say, on a component of (6). The solution space of (14) is spanned by the spinor fields ω^{AB} of the form[10]

$$\omega^{AB} = \omega_1^{(A}\omega_2^{B)} = \omega^A\omega^B \qquad (15)$$

where each ω_i^A is a valence-1 twistor satisfying the equation

$$\nabla_{A'}^{A}\omega^B = -i \in^{AB} \pi_{A'} \qquad (16)$$

In all the times of our measurements, we have the conservation condition given for the equation

$$\nabla^\alpha T^{\alpha\beta} = 0 \qquad (17)$$

which is to say, we suppose that the energy-matter is always present in the space and is constant, at least in the space-region bounded by the 3-dimensional surface Σ. If a supermassive body that perturbs space-time exists, the energy-matter of its tensor can be carried to (see fig. 11)

$$A_{\alpha\beta} = \frac{1}{16\pi G}\oint_S R_{AB}\omega_\alpha^A\omega_\beta^B \qquad (18)$$

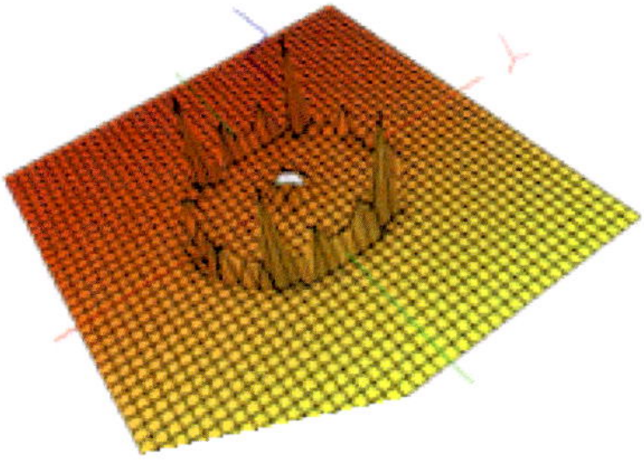

[10] Here, the product of the spinors $\omega_1^{(A}\omega_2^{B)}$ comes from product of the fields $Z_1^{(A}Z_2^{B)}$, which is a symmetric tensor product. That is to say $Z_1^{(A}Z_2^{B)} = Z_1 \otimes_{\text{Symm}} Z_2 \in \mathbb{T} \otimes_{\text{Symm}} \mathbb{T} = \mathbb{T} \odot \mathbb{T}$.

Fig. 11. Kinematic twistor tensor due to energy-matter tensor perturbation of the supermassive body, determined on the sphere *s*.

Finally, we can establish the following commutative diagram of mappings of twistor spaces on the gauging and detection mechanism of the torsion

$$\begin{array}{ccccc} \mathbb{T}(\Sigma) & \xleftarrow{I^{\alpha\beta}\Sigma_{\beta\beta'}} & \Sigma & \xrightarrow{A_{\alpha\beta}} & (\mathbb{T}(S)\odot\mathbb{T}(S))^* \\ I_{\Sigma}^{\alpha\beta}\uparrow & & \uparrow \mathrm{T}^{\alpha\beta} & & \uparrow A_{\alpha\beta}Z^{\alpha}Z^{\beta} \\ \mathbb{T}(S) & \xleftarrow{I^{\alpha\beta}} & S & \xrightarrow{\omega^{AB}} & \mathbb{T}(S)\odot\mathbb{T}(S) \end{array} \quad (19)$$

where $\odot$ is the symmetric tensor product.

2.3. Curvature Energy and Torsion

The following results, obtained in [1], present the fundamental principle that is required to gauge and detect the torsion of the tensor $A_{\alpha\beta}$ considering the law of transformation to pass from field Z to Z^{β} to two coordinate systems α and β and to transform the surface Σ through the transformation law (as in diagram (19))

$$\Sigma_{\alpha\beta} = A_{\alpha\beta} I^{\beta\gamma} \Sigma_{\gamma\alpha'} \quad (20)$$

Thus, we can state the following theorem.

Theorem 2.3.1 (F. Bulnes, Y. Stropovsvky, and I. Rabinovich). We consider the embedding of

$$\sigma : \Sigma \to (\mathbb{T}(S) \otimes \mathbb{T}(S))^* \qquad (21)$$

The space $\sigma(\Sigma)$ is smoothly embedded in the twistor space $(\mathbb{T}(S) \otimes \mathbb{T}(S))^*$ and their curvature energy is the energy given in the interval $M_N \geq A_{\alpha\beta} Z^{\alpha} I^{\beta\gamma} \overline{Z}_{\gamma} \geq 0$.

Proof **[1]**

We have a source for the linearized gravitational field that is explained through the kinematics and electrodynamics used in its construction (see fig. 13). The correspondence of the linearized Riemann tensor to the spinor frame has been constructed considering the components

$$f_{\alpha\beta} = \omega_{AB} \in_{A'B'}, \qquad (25)$$

which relates the spinor field ω_{AB} to the Killing vector k^{a} in the twistor equation for a valence-2 twistor. We use the divergence theorem when S is a 2-surface in the 3-surface Σ, given as

$$\Sigma = \omega^{A} \overline{\pi}_{A'} + \omega^{A} \overline{\pi}_{A'}, \qquad (26)$$

around the source and having several censorship conditions determined by the dominant energy conditions of the curvature that can be used in electronic experiments.

We have a metrology [11, 16, 17] of the curvature measured and detected by our curvature sensors and that permits us to find the curvature in new units obtained from a strong electronics-gauging study [11, 14].

Likewise, the energy of the kinematic twistor tensor will be the measurement of the substantive energy in relation to the curvature energy in the case of the spinor-twistor frame given in the energy domain $M_N \geq A_{\alpha\beta} Z^{\alpha} I^{\beta\gamma} \overline{Z}_{\gamma} \geq 0$.

The solution of the quasi-local mass is directly related to the quantity of the energy-matter tensor. This solution is a function of the radius and time, as a wave-pulse, which can be spectrally reproduced in the function $\frac{\sin \omega L}{\omega L}$, with the voltage of the electronic device of the electromagnetic radiation bath interacting with the proximity of the supermassive object or simple mass in motion (see figs. 12, 13 A), B)).

A)

VOLTAGE	Length
1.62	5
1.62	4.5
1.62	4
1.63	3.5
1.64	3
1.64	2.5
1.65	2
1.66	1.5
1.68	1
1.7	0.5
1.8	0

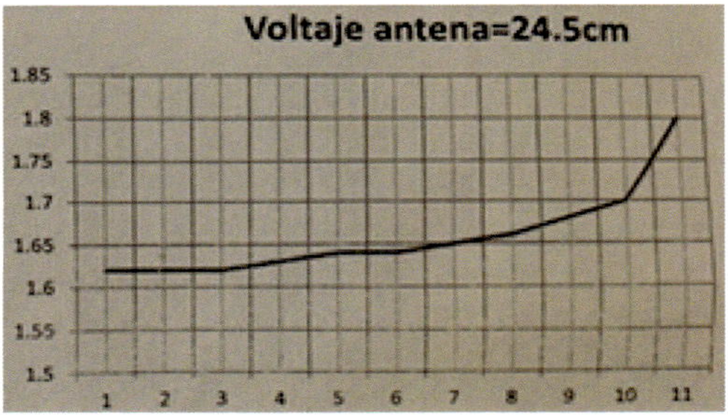

B)

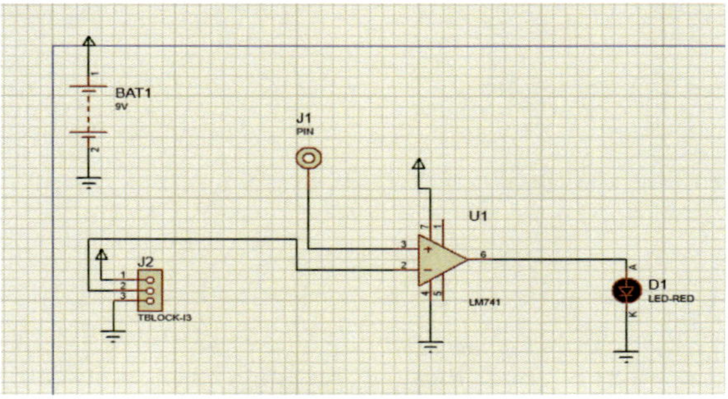

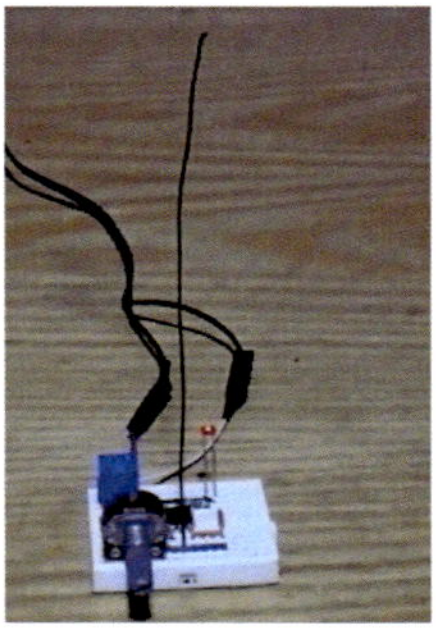

Fig. 12. A) Antenna with voltage feeding of length 24.5 cm. B) Electronic device with electronic monopole connected to electromagnetic radiation bath.

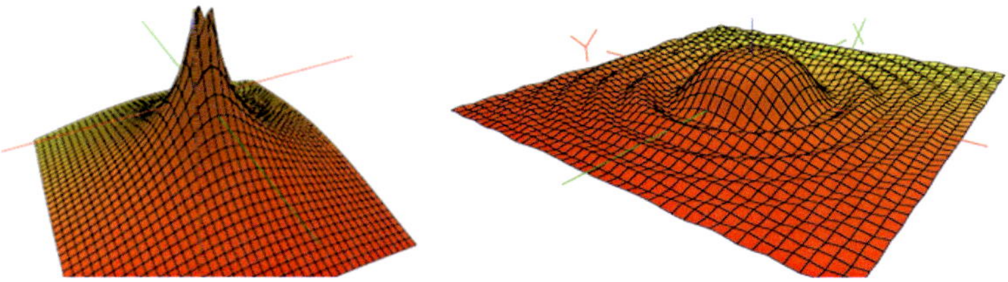

A) B)

Fig. 13. A) Monopcle field with a 2-D surface of charge $Q[k]$. B) Supermassive body 2-D surface of energy-matter tensor $T^{\alpha\beta}$.

2.4. Electronics Experiments to Demonstrate the Existence of Torsion Through Wave-links as Spinors and Wave-pulses

An electromagnetic field as a detector can also help establish a perturbation in space-time allowing us to perceive the existence of torsion. Furthermore, this field must comply to a solution of the Maxwell equations in the spinor-twistor frame such that[11]

$$\phi_{A'B'\ldots L'}(\mathbb{R}^{a})=\frac{1}{2\pi i}\oint_{Z^{a}\leftrightarrow R^{a}}\pi_{A'}\cdots\pi_{L'}f(Z)\pi_{F'}d\pi_{F'} \qquad (27)$$

and

$$\phi_{A'B'\ldots L'}(\mathbb{R}^{a})=\frac{1}{2\pi i}\oint_{Z^{a}\leftrightarrow R^{a}}\frac{\partial}{\partial\omega^{A}}\cdots\frac{\partial}{\partial\omega^{L}}f(Z)\pi_{F'}d\pi_{F'} \qquad (28)$$

For the particular case of the determination of $A_{\alpha\beta}$, we have the integrals

[11] Here, the electromagnetic wave equation can be characterized by the massless field equations $\nabla^{AA'}\phi_{AB\ldots L}=0,\quad \nabla^{AA'}\phi_{A'B'\ldots L'}=0$, which are equivalent to $\Box\phi=0$ for the case of zero spin.

$$
\begin{aligned}
\mathrm{A}_{\alpha\beta} &= \frac{1}{16\pi G}\oint_{S} R_{AB}\,\omega_{\alpha}^{A}\omega_{\beta}^{B} \\
&= \frac{1}{16\pi G}\oint_{S} R_{\alpha\beta\gamma\delta} f^{\alpha\beta} d\sigma^{\gamma\delta} \\
&= \frac{1}{2\pi i}\frac{i}{8\pi G}\oint_{S} \omega_{\alpha}^{A} d^{2}\omega_{\beta A} \\
&= \frac{1}{16\pi G}\oint_{S} \omega_{\alpha}^{A} d^{2}\omega_{\beta A}
\end{aligned}
$$

where the field value around the circle used has been applied as a cycle of displacement along a cylindrical spiral for three cycles (see fig. 14). As discussed in section 2, the evidence for torsion can be obtained with good approximation (given the limitations of the electronic system) when a complete signal $\frac{\sin \omega L}{\omega L}$ is obtained for each three cycles producing two complete spinors.

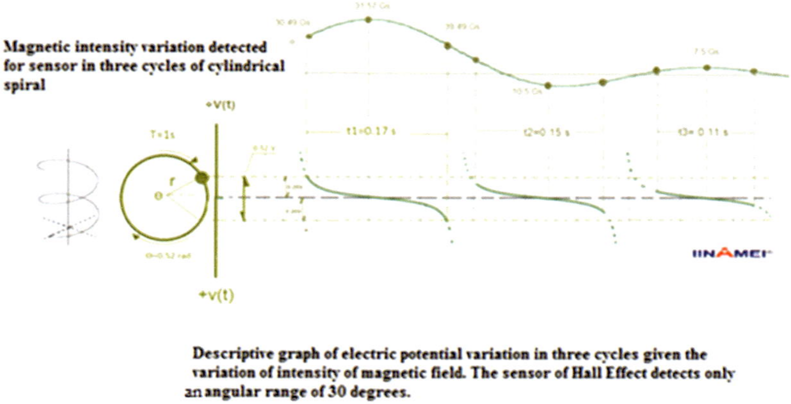

Fig. 14. Dynamic-magnetic system to define $A_{\alpha\beta} Z^{\alpha} I^{\beta\gamma} \overline{Z}_{\gamma}$.

The sensor device for the magnetic field of the Hall effect is able to detect the boundary region with an arc length of 0.045 *m* (see fig. 15 A)). Without this range, we cannot detect the field although the subsequent cyclical displacements of the magnetic dilaton. This is shown by the three curves in the graph in figure 14, with the displacement times t_1, t_2, and t_3. The electric potential that is generated by the variation of the magnetic field is inversely proportional to the magnetic field intensity with a base in the relation 1.4 *mV/Gs*.

In the first half of the process, the magnetic dilaton generates a decreasing potential of 0.52 *V* to a minimum of 0.26 *V*. In the second half, the magnetic dilaton generates increasing potential of 0.26 *V* to a maximum of 0.52 V when it moves away. For subsequent cycles, the remoteness of the sensor in the trajectory obeys the spiral trajectory of the system dynamics. The effect of the magnetic dilaton and the dynamics of the system define the twistor kinematics tensor $A_{\alpha\beta}$, which can be gauged in a finer way using a quantum electronic device version of the electronic system used in this experiment. The energy-mass tensor depends on the gravitational field between the dilaton's mass and the Earth's mass. The coordinate systems $A, B, \ldots, L,$ and $A', B', \ldots, L',$ are considered the inertial reference frames used in the experiment.

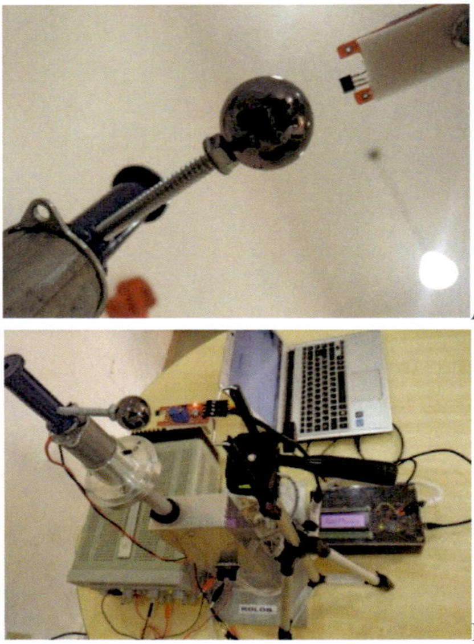

A)

B)

Fig. 15. A) A magnetic sphere as a magnetic dilaton of radius 0.025 *m*. This dilaton is used to detect distortion in the boundary surface Σ, which results from the interaction between the magnetic field of the proper dilaton and the gravitational field generated for the mass of the proper dilaton and the mass of the Earth. The maximum proximity of the sensor is 0.010 *m*. B) A rotational dynamic system of radius 0.085 with reversible vertical displacement of 0.040 *m*. The sensor used is a Hall effect sensor. The device has an interface system for a microcontroller and symmetric variable voltage source.

The conditioning of the signal is defined for continuous variation of the electrical potential with frequency conversion through the LM331 integrated circuit. The maximum response (frequency output) of this device is 10 *KHz* and thus we have developed an electronic circuit to condition the signal. The digital signal obtained for each electrical potential variation (0.52 *V* to 0.26 *V* and 0.26 *V* to 0.52 *V*), as a result of the change in position of the magnetic dilaton in the space, is established. The motivation for having a digital signal with appropriate pulse width for the 26 respective positions in space is so that we can do a convolution with a sinosoidal signal for each pulse and thus obtain periodic signals for the points studied, determining the curve in 3-dimensional space in field theory in terms of signal analysis.

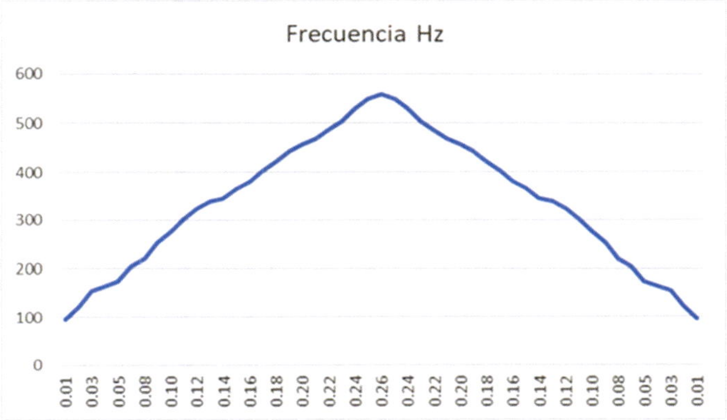

Fig. 16. Frequency in the trajectory of the first cycle.

In the first experiment (described in section 2. 3), the sphere *S* does not have a curved 3-dimensional surface Σ. The electromagnetic field of the monopole is fixed and does not produce a distortion of space. All particles of matter complying with this spherical symmetry fall in the natural gravitational field of Earth.

In the second experiment (in section 2. 4), the chosen magnetic dilaton is a ball of certain mass, which is displaced along a cylindrical-spiral trajectory producing a distortion, measurable by electronic devices, in space, which can be affected by the Earth's magnetic field and the gravitational field between the dilaton's mass and the Earth's mass. Before we consider the following 2-dimensional surface model of the spinors, we directly deduce from the second experiment and verify some of the conclusions about the torsion's existence and the twistors (see fig. 17).

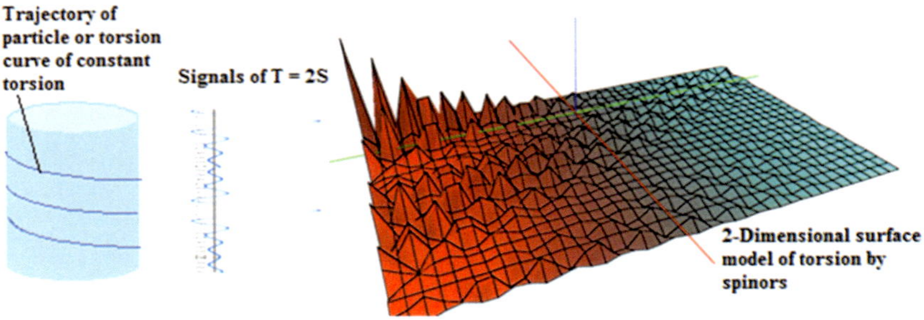

Fig. 17. 2-D model of torsion with spinors.

We can establish different dualities in field theory concerning geometry and movement to relate the energy-matter tensor to the kinematic twistor tensor in the study of torsion. A torsion is a field observable, or, in geometry, a second curvature. From the point of view of field theory, torsion presents significant evidence of the birth of gravity and its consequences up to now with gravitational waves detected by astronomical observatories.

We can design an electronic analogue for the measurement of torsion as evidence of the existence of gravitational waves and perform an experiment to give some insight into the study of theories of gravitation; however, with modern mathematical study into invariants, such as twistors and spinors, we operate at the microscopic level and use microscopic field theory.

The limitations of purely electronic devices mean that we can only see certain traces of electronic signals and interpret the relevant data using arguments of geometry to formulate the evidence for torsion. We consider an eletromagnetic field determined in a specified voltage range and movement along a cylindrical trajectory, which as we know, is a constant torsion. This verifies conjecture 2.1.1 and the theorem in 3.1 [1], which have been established in other studies in the fields of theoretical physics and mathematical physics. Likewise, the methods and results of this reseach concern themes related to the concept of gravity (although not precisely so) and consider this method as analogous to the detection gravity waves—in this case, indirectly detecting torsion waves.

2.5. The Torsion of Gravitational Action: Some Digressions

Einstein presented the gravitational field equation $R^{\mu\nu} - \frac{1}{2} Rg^{\mu\nu} = -\chi T^{\mu\nu}$, which can be simplified to $R + \chi T = 0.$ This is the basic postulate of general relativity. The constant χ is the gravitational constant given by Einstein and equal to $8\pi G / c^2$.

As we know, *R* is a measure of curvature in differential geometry (for example, κ^2), and *T* is a measure of energy density (mass, pressure, self-gravitation or auto-gravitation, and/or 4–velocity). In the limits of a weak field $\chi, T = m/V$, that is to say, the mass by volume is equal to the density.

Following on from relativity, the equation $R + \chi T = 0$ establishes a curvature in the presence of energy in space-time experiments. That is to say

$$R = -\chi T (29)$$

The negative sign in $-\chi T$, is a convention, but it can serve to remind us that the energy curves of space-time are inside.

Thus, we can see gravity "winging" the things to their source. A little change in the paradigm here is that the mass is not attracted between these "wings" in space-time, but rather curves to space-time.

An important observation is that when using χT, we must imagine a spherical density. In changing this, when we find *R*, the curvature is in space-time as a deformation of the spatial stage due the presence of matter. The exact values are not important in this last description.

In non-Euclidean space-time or Riemannian space, the symmetric tensor of the metric is $o^{\mu\nu(S)}$. This is the same as *g* in the metric tensor, but referring to distances that are functionally symmetric. Here, the superfix *S* means it is symmetric.

The asymmetric tensor of the more general metric is defined through the external product or tensor between tetrads (four metric vectors), that is to say

$$o_{\mu\nu}^{ab} = o_{\mu}^{a} o_{\nu}^{b} = o_{\mu\nu}^{ab\,(S)} + o_{\mu\nu}^{ab\,(A)} \qquad (30)$$

The asymmetric form has internal symmetry, as well as anti-symmetry. The anti-symmetric component from (30) is $o_{\mu\nu}^{ab\,(A)}$. The anti-symmetric tensor of the metric is defined through the wedge product of two tetrads.

$$o_{\mu\nu}^{ab\,(A)} = o_{\mu}^{a} \wedge o_{\nu}^{b} \qquad (31)$$

Due to the fact that symmetry is defined by the existence of centralized potentials, that is to say, to spherical forms (as in gluing), the anti-symmetrical apsect always involves rotational potentials and the asymmetry indicates that both are included inside the same form; in geometry, distance implies spin. This means that, in certain forms, the anti-symmetric and symmetric forms are perpendicular. The symmetric vectors give us distances in 4-dimensional space-time, while the anti-symmetrical vectors give us spin effects of space-time (see fig. 18). Thus, the tensor gravity complies with

$$G^{\mu\nu} = G^{\mu\nu(S)} + G^{\mu\nu(A)} \qquad (32)$$

Considering again the Einstein equation for gravitation, but now in the shape given by the tetrads and the metric, we write

$$R_{\mu}^{a} - \frac{1}{2} R o_{\mu}^{a} = \chi T_{\mu}^{a} \qquad (33)$$

Equation (33) gives us distances in space-time that define the curvature. From the basic structure of equation (33), we obtain three types of field equations

$$g_{\mu\nu}^{(S)} = o_{\mu}^{a} o_{\nu}^{b} \eta_{ab} \qquad (34)$$

$$g_{\mu\nu}^{ab(A)} = o_{\mu}^{a} \wedge o_{\nu}^{b} \qquad (35)$$

$$g_{\mu\nu}^{ab} = o_{\mu}^{a} o_{\nu}^{b} \qquad (36)$$

where S denotes symmetry and defines distance and A denotes anti-symmetry and defines *torsion*. In equation (36), the tensor $g_{\mu\nu}^{ab}$ is the combined asymmetric metric. This form is simultaneously symmetric and anti-symmetric.

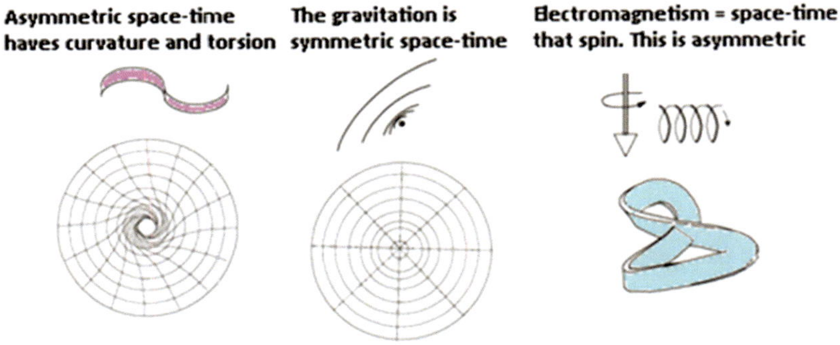

Fig. 18. 2-D models of universes with anti-symmetric, asymmetric, and symmetric forms. The asymmetric space-time is "glued" *(adjunction)* by the two components of tensor $G^{\mu\nu}$ through $X +_f Y$[12].

[12] The "gluing" effect of gravitation generates loops and sedimentation by the accumulation of matter in different intensities of the gravitational field. All bodies are linked or looped with others through gravity. This is field gravity and every body is a source of this gravitation, that is to say, every body is "winging" with another and vice versa. The linkage join the layers of different gravitational intensity satisfying the logarithmic increase in the curves. Then, the set of these links and loops that transmit the effect gravity from layer to layer is a loop space with loops or links formatted by certain wrappings and strings at the quantum level. The addition of gravitational energy through the *gluing of particles* and *quantum elements* produces *matter*. We will describe these ideas more precisely. An *adjunction* (*or gluing*) *space* (or *attaching space*) is a common construction in topology where one topological space is attached or "glued" to another. Specifically,

Using equation (35), we at a new field equation of electrodynamics that is generally covariant

$$o_{\mu}^{a} \wedge (R_{\nu}^{b} - \frac{1}{2} R o_{\nu}^{b}) = \chi o_{\mu}^{a} \wedge T_{\nu}^{b} \qquad (37)$$

which give us the spin or torsion of the field. We can use an extended Einstein theory, for example the Cartan-Enistein theory, to explain how the gravitational equation of Einstein can be deduced from (37) and also from the electromagnetic equation (35) through the use of the wedge product.

Using this asymmetric space-time model, we obtain field models that reflect torsion, for example the asymmetric field theory of Yang-Mills, which provides the extension of Maxwell's theory to the case of non-Abelian fields. In this dimension, we raise the wrappings and loop contributions that contribute to the energy micro-states used to define entropy in atomism (see fig. 19).

---

let X and Y be topological spaces with V as a subspace of Y. Let $f: V \rightarrow X$ be a continuous map (called the *attaching or gluing map*). The glued space $X \bigcup_f Y$ is formed by taking the disjoint union of X and Y, identifying x with $f(x)$ for all x in V. Schematically, $X \bigcup_f Y = (X \coprod Y)/\{f(V) \triangleright V\}$.
Sometimes, the adjunction is written as $X +_f Y$. Intuitively, we think of Y as being glued onto X via map f. As a set, $X \bigcup_f Y$ consists of the disjoint union of X and ($Y - V$). The topology, however, is specified by the quotient construction. In the case where V is a closed subspace of Y, it can be shown that the map $X \rightarrow X \bigcup_f Y$ is a closed embedding and $(Y - V) \rightarrow X \bigcup_f Y$ is an open embedding.
If we consider that the adjunction space is given when Y is a closed n-ball (or *cell*) and V is the boundary of the ball, the (n–1)-sphere inductively attaches cells along the spherical boundaries of this space, resulting in an example of a *CW*-complex, which is homotopically useful.

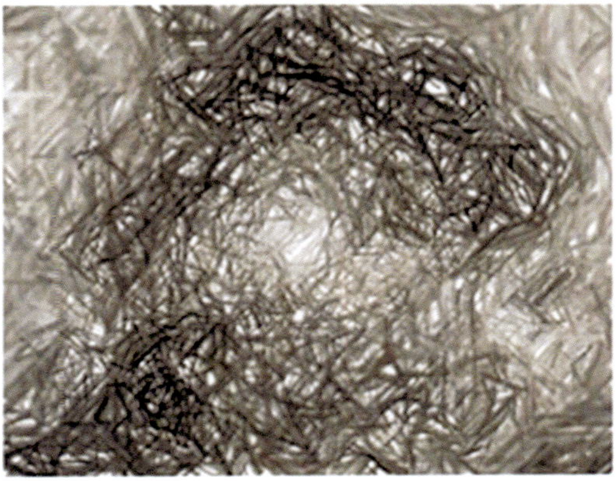

Fig. 19. The micro-states.

Using the the Cartan tetrads $O^{ab}_{\mu\nu}$ we can obtain a quantum field equation (*tetradic wave equation*) that is analogous to the Dirac equation, considering the electromagnetic and gravitational fields

$$(\Box + \chi T) o^{a}_{\mu} = 0 \qquad (38)$$

This *alternative theory of field unification* could permit the definition, with precision, of the mechanism between gravity and electromagnetism. Using *homotopies*, *operads*, and *certain algebras* we can define the quantum mechanism of field ramifications that determine movement. Furthermore, this link between the field theories at the macroscopic and microscopic levels explains the gluing models of field gravity. But what happens with entropy?

The reality is that there is no entropy in our scheme of mirror space. All the interchanges of energy that begin from the factor in (38) are given by χT.

Then, using equations (33) and (38) all energy-momentum is interchangeable, and the corresponding micro-states obey, by an order of magnitude, a more probable arrangement of particles and strings. Thus, the production of energy and matter can be written out for equations (33) and (38) as

$$T = T_{\text{gravity}} + T_{\text{ele/mag}} + \ldots + T_{\text{hybrid}} \qquad (39)$$

Entropy is an elegant and prominent concept of thermodynamics and its use in this description is based on the expression of the macroscopic properties of its system, revealing the behaviour of the primary and microscopic structures of matter. In composing systems of many parts, this is the sum that exists in every one of them.

If a transformation happens in an insolated system, the entropy of the final state cannot be less than the initial one, since these do not acquire free energy. Insolated systems of maximum entropy, that is to say, with minimal free energy, cannot change, since any other transformation could result in a decrease in entropy, which is impossible. Thus, we can establish that, for an insolated system, the state of maximum entropy is the most stable one possible: this will be the case of the universe in the distant future.

We can se the adjunction of energies in equation (39) with a function of the micro-states of space and some of these energies are shown in (39). Likewise, the entropy ***S*** is the function of the space $\mathcal{H}$ of the micro-states $p_i (i = 1,2,3,\ldots,n)$ with values in some of the manifolds that model the universe[13]. For example, all the scalar curvatures are interchangeable in form (as reflected in the structure $X +_f Y$!)

$$R = R_{\text{gravity}} + R_{\text{ele/mag}} + \ldots + R_{\text{hybrid}} \qquad (40)$$

and these define the geometry of a single universe[14] [18].

---

[13] Our function in this case is $s : \mathcal{H} \to X$.

[14] Boltzmann related the second law of the thermodynamics with the atomist image of the universe. He showed that entropy is a function of a thermodynamic macro-state that happens to agree with the realization of the micro-states, that is to say, describing entropy as a measure of molecular disorder and related to the probability concept and studies of statistical mechanics. According to the work of Boltzmann, the entropy of a system increases in relation to how the molecular structure system acquires a more probable arrangement. Likewise, the entropy of the set can be defined when the fraction of the system that corresponds to every micro-state is

Now we may ask, what is the relation of entropy to torsion?

Studying the properties of Killing horizons in the presence of torsion, we can establish the notion of a local causal horizon in Riemann-Cartan space-time and derive the generalised Raychaudhuri equation[15] for this kind of geometry. Then, starting with the entropy that can be associated with these local causal horizons, we derive the Einstein-Cartan equation by implementing the Clausius equation. We can take torsion as a geometric field or as a matter field. In both cases we need to add internal entropy production terms to the Clausius equation, since the shear and twist cannot be taken to be zero a priori for configuration in this case. This fact implies the necessity of a non-equilibrium thermodynamics for the local causal horizon. Thus, torsion is linked to entropy in a thermodynamic equilibrium in space-time. However, this does not happen in reality because of the

---

known. Thus, a thermodynamic micro-state exists that agrees with the realization of the micro-states

$$s = -\xi \sum_i p_i \ln p_i$$

which is the formula that relates entropy S to the Boltzmann constant $(\xi = 1.3806504 \times 10^{-34}\, J/K)$ and the corresponding fraction to the *ith*-micro-state p_i. If the set is distributed in a given number of micro-states of *m*-probability such that they are equally possible, then all P_i will have the same value

$$s = -\xi m \frac{1}{m} \ln \frac{1}{m}$$

and the entropy can have its mechanics explained (statistical mechanics) by

$$s = \xi \ln m$$

where m is the probability of the state and k is the Boltzmann constant. The number of m that can risk the state of the constituted system, for example, to two bodies, is equal to the sum of the numbers m_1 and m_2, thus

$$s = \xi \ln m = \xi \ln m_1 + \xi \ln m_2 = s_1 + s_2$$

conserving the energies from the field given in equation (10).

[15] The equation for extending relativity theory established by Cartan-Einstein theory is important as it is a fundamental lemma for the Penrose–Hawking singularity theorems and for the study of exact solutions of general relativity. It also is of independent interest, since it offers a simple and general validation of our intuitive expectation that gravity should be a universal attractive force between any two bits of mass-energy in general relativity, as is the case in Newton's theory of gravity.

existence of different material components of the universe that have been "decoupled" at different times. For example, fairly early on (a few minutes after the Big Bang), neutrinos stopped interacting significantly with anything else. This was due to both the density of the universe dropping (resulting in fewer "targets" for the neutrinos) and to the temperature dropping (lower-energy neutrinos react less easily than higher-energy neutrinos). Dark matter was most likely thermally decoupled very early on in the development of the universe. Dark matter may be very much related to torsion [9], for example, in the various steps of the early universe's development, as with baryogenesis and leptogenesis.

2.6. Gravitational Energy and Movement

We consider here some aspects related to movement as consequence of energy-vacuum interactions that can explain the thermodynamic non-equlibrium state generated in the early or incipient step of the universe's development immediately after the Bing Bang and the irregular concentration of incipient matter as baryons. The symmetry of this incipient matter was broken resulting in high quantities of ordinary residual matter in the universe today, although big extensions of vacuum potentially produce a vacumm effect (*expansion*) linked to torsion.

Def. 2.6.1 (F. Bulnes, M. Ramírez, L. Ramirez, O. Ramírez). A *broson* is a hypothetical particle constructed from fermions coming from *D*-branes and wrapped by gauge bosons in space-time. In a certain sense, the *broson* offers a solution to equation (38), defining other tools and applying other theories.

The gauge bosons produce torsion in microscopic space due their electromagnetic characteristics as photons realizing back-reactions in the space affected by gravity. In fact, in non-Abelian electromagnetic theory, we find Faddaeev-Popov "ghosts" (states of negative norms or fields with the wrong sign of the kinetic term) linked to every particle whose effect can be predicted by the Faddaeev-Popov procedure.

Every gauge field has an associated ghost and where the gauge field acquires mass via the Higgs mechanism, the associated ghost field acquires the same mass (in the *Feynman-'t Hooft gauge* only; it is not true for other gauges).

From the point of view of QFT and particle physics, there is a weak field that defines the integration of the neutron, which is associated with the gauge boson, that is to say, the W boson. This weak field has an electromagnetic nature. The neutron, when it exists outside the atomic nucleus, is transformed after ten minutes or so to an electron, an anti-neutrino, and one proton. As such, it is ready to shape matter and antimatter in the required proportion to render the material image of our universe.

A substantial requirement in this movement is the quantity of energy involved. For us, movement is the displacement of a thing in space; however, movement always involves the modification of an object or thing. This modification naturally requires energy to occur and become a reality. All sensitive things have movement as one of their traits defining further characteristics.

Now, we express movement as a continuous transformation that comes from a field and this has an algebraic object assigned, as a set of rules of a chain complex (as a graded-module equipped with degree W such that $d^2 = 0$) that permit their development in a quantum context through deformation of space-time at the microscopic level. The macroscopic appearance of such quantization can be consigned inside a family of Poisson manifolds, which, under the influence of certain quasi-equivalences [19] and homotopic classes, can be carried to a macroscopic reality. Defining the duality of Koszul complexes in relation to micro-state spaces, we can demonstrate that entropy is an aspect of the evolution of energy. We can consider it as an inverted image in mirror space of the equivalences presented according to the algebraic context and with respect to the quantization process. Gravity, in this case, is consigned to torsion (and defined as pressure on a body or particles) in a *Drinfeld space* with *twisted loops*, where these loops and strings could be our brosons according to the definition given in 2.6.1.

The torsion model must lie in equation (38) and the explicit values can be obtained by integrals (*solutions* to (38)) developed in the microscopic model [rusos]. Then, as the curvature sees immediate effect (equal to the gravitational energy) from the gravitational field, with gravity being the mother cause of this model, we call it a *ramified gravitational field* where time is the distance between cause and effect. Thus, the curvature in this case can be viewed as time defined by the *energy curvature states* up to the finite actions of the field. Every lapse between the layers of gluing in the gravity model is a time in Minkowski space (*invisible yarns of wires*). This is our first conjecture of *time and gravitational energy*. Movement is obtained when more layers are added, that is to say, gravity has ramifications as a consequence of the trajectory of the electron moving to different times.

We note here that the term "gravitational ramified field" concerns the gravity that results from the addition of layers, in which every layer is of a certain dimension such that their sum shapes a space of certain co-dimensions in respect to Einstein space-time. In general, given the shape of the field equations, an instanton[16] (defined by minimal paths of maximum benefit) defined in a Euclidean space of N, has dimensions that are equivalent to a static solution of the same theory, defined in a space-time of $N+1$ dimensions. The missing dimension is time from the macroscopic perspective. The instanton uses a radial transformation on classic path integrals and represents the quantum displacement *for one path of major statistical weight*.

Thus, the curvature energy is the displacement consigned through the Cartan tensor $O_{\mu\nu}^{ab}$ and then the variation of the curvature energy is their torsion energy, possibly from the velocity of the mass flux that appears in the Lorentz-invariant theory of gravitation (see fig. 20).

---

[16] Mathematically, a *Yang-Mills instanton* is a self-dual or anti-self-dual connection in a principal bundle over a 4-D Riemannian manifold, which plays the role of physical space-time, as mentioned, in non-Abelian gauge theory. Instantons are topologically nontrivial solutions of the Yang-Mills equations that absolutely minimize the energy functional within their topological type. The first of these solutions were discovered in the case of 4-D Euclidean space compactified to a 4-D sphere, and turned out to be localized in space-time, prompting the names *pseudoparticle* and *instanton*.

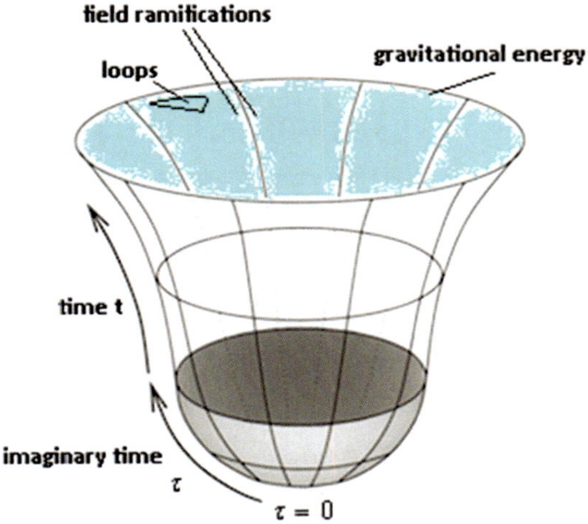

Fig. 20. Instanton of imaginary time τ.

CHAPTER 3

INTERACTIONS BETWEEN THE FERMIONS OF MANY FERMION GENERATIONS

3.1. From the Big Bang up to Today's Universe

As has been mentioned previously, the torsion is a field effect that arose during the Big Bang and has passed across different epochs and stages. Its nature has been determined by the processes of leptogenesis and baryogenesis, as well as different neutrino species and different fermions (fermiongenesis), leading to the creation of the first heavy particles and the formation of matter as hadrons.

From this we derive the mechanism of the cohesion of dark energy that results in dark matter and gravitation, with additional consideration of photons, in the conformation of material particles (electrons, protons, and neutrons) and their transformation in the atom—this last aspect being a fundamental part of matter. Basically, this step has resulted in the abundant proliferation of hadrons and neutrinos giving us the universe that we know today.

In this chapter, we study the field torsion mechanism that creates matter along with their corresponding Dirac equations for neutrinos and matter/anti-matter asymmetries, gravitational insights, and going beyond neutrinos. We conclude that the role of the field torsion has been relevant at all times and fundamental, even, to the creation of life.

Observing the predominance of matter over anti-matter in the actual universe, with the creation of matter established through mechanisms that utilize the smallness of the mass of active neutrino species, we can define the material universe as a space-time at the quantum level that has been created through these fragments. This involves the breaking of heavy particles such as baryons that conform to asymmetries that may explain the proliferation of leptons in the polarizable constitutive elements of neutrinos resulting in quantum torsion. Likewise, this quantum torsion can defined by Majorana neutrino masses, from which dark matter is formed as an ulterior product (see fig. 21).

Fig. 21. Evidence of dark matter in the observable universe. This sidereal object is a galaxy group in the Canis Major constellation. The very bright light is due to high concentrations of sterile neutrinos.

One aspect that propitiates this conclusion is *KR*-background torsion, which yields matter/anti-matter asymmetry using gauge-field frames. This means that we can use the bosonic *KR*-torsion from string theory to generate the matter/anti-matter asymmetry in a string model of the universe. Use of these boson fields also leads us to use pseudo-scalar fields (as a new version of scalar field theory with infinite extensions of the Standard Model (SM)), as being responsible for both right-handed Majorana neutrino masses and the homogeneity of the Dirac equations without the need for extra adjustment given by the *CP*-violation. In addition, we have two Dirac equations corresponding to two Majorana neutrino masses and their relation shapes the quantum torsion as we wish.

3.2. Torsion: Spectra and Interactions Between Fermions of Many Generations

In Einstein's theories (general and special relativity), the torsion was absent of constraint as a field observable being rather an effect that was seen to depend on matter/anti-matter interactions, which not have been implied in the model of torsion as a dynamic field (as a combination of two fields: the electromagnetic and the gravitational). It is necessary to consider that, since torsion is not independent of the vierbeins[17] (or Cartan connection [20]) at the fundamental level (particles), fermions appear in a gravitational field through a first-order formalism of the spin connection and veirbeins, treated as independent, and the torsion does not vanish.

The torsion, as an invariant of energy-space, remains defined as

$$T^{\gamma}_{\alpha\beta}=2S^{\gamma}_{\alpha\beta} \qquad (41$$

The corresponding spinor representation of the spectral torsion considering (41) is

$$\tau(\omega_1,\omega_2) = 2s(\omega_1,\omega_2), \qquad (42)$$

However, the expression of spinor spectra $s(\omega_1,\omega_2)$ is not easy to obtain because it requires the microscopic treatment of the behavior of the quantum particles that are sources of torsion. This treatment could seem to act on a "plane" in some aspects, but on $S^{\gamma}_{\alpha\beta}$, and the current density J^{α}, it stays linked in the expression, possibly of a major order[18], through which we can

---

[17] The vierbein or tetrad theory used in theoretical physics is a special case of the application of a Cartan connection in 4-D manifolds. It can be applied to metrics of any signature.

[18] The microscopic torsion theory involving the solution of the field equations

$$\Im_G + \Im_{\text{QED-fermions}} = \frac{1}{2\chi}\int d^4x o o^a_\mu o^b_\nu R^{\mu\nu}_{ab}(\omega) +$$

$$\frac{i}{2}\int d^4x o(\bar{\psi}\gamma^\mu \mathcal{D}_\mu(\omega,A)\psi - \overline{\mathcal{D}_\mu(\omega,A)\psi}\gamma^\mu\psi) + \ldots$$

establish the possible sources of torsion [21]. Analysis of the interactions between fermions of many generations (dispersion) can be use to determine, or at least prove the appearance of torsion on the microscopic scale. In relation to torsion phenomena, there are many aspects concerning the effects that result from interactions between fermions and the role of neutrino production in the beginning of the universe. Such as can be viewed, in the description of gravitational phenomena, cannot be disconnected from the torsion and the proper fermionic nature of the phenomena that resulted from the Big Bang. As such, the oscillations of the spinors can be used to generate descriptions of these phenomena as the mixture of dilaton oscillations and neutrinos/anti-neutrinos can be measured through the spinors. However, we need to establish the front of the "waves" at this scale [22].

---

to the Dirac matrices $\{\gamma^{\mu},\gamma^{\nu}\}=2g^{\mu\nu}$ and where $R^{\mu\nu}=d\omega^{\mu\nu}+\omega^{\mu}_{\sigma}\wedge\omega^{\nu\varsigma}$ is the 2-form of the curvature. Also $\chi=8\pi G_N$, where $G_N=G/c$ and C is the velocity of light in special relativity. G is the Newtonian constant of universal gravitation.

CHAPTER 4

TORSION AND THE CORRESPONDING DIRAC EQUATIONS OF NEUTRINOS AND MATTER/ANTI-MATTER ASYMMETRIES: GRAVITATIONAL INSIGHTS

4.1. Introduction

We consider the Cartan-Einstein theories in analyzing the aspect of torsion relevant to the dynamic field, as torsion is not independent from the vierbeins (or Cartan connection [23]). This can be explained by the relations between gravity and torsion, which are linked by gravity, and the asymmetric connections established in *gravity with torsion*, with the connection given by

$$\omega_{\mu}^{ab} = \overline{\omega}_{\mu}^{ab} + K_{\mu}^{ab} \qquad (51)$$

where the connection $\overline{\omega}_{\mu}^{ab}$ is given as

$$\overline{\omega}_{\mu}^{ab} = e_{\nu}^{a}\partial_{\mu}e^{\nu b} + e_{\nu}^{a}e^{\sigma b}\Gamma_{\sigma\mu}^{\nu} = e_{\nu}^{a}e_{\mu}^{\nu b} \qquad (52)$$

and the other connection is given by K_{μ}^{ab}

$$K_{\mu}^{ab} = e_{\nu}^{a}e_{\rho}^{b}K_{\mu}^{\nu\rho} \qquad (53)$$

such that

$$K_{\mu}^{\nu\rho} = -K_{\mu}^{\rho\nu} \qquad (54)$$

where K_{μ}^{ab} is the *contorsion tensor*.

For the other side, the covariant derivative of the fermions is

$$D_{\mu}\psi = \left(\partial_{\mu} - \frac{i}{4}\omega_{\mu}^{ab}\sigma_{ab}\right)\psi \qquad (55)$$

with the tensors from the Dirac matrices $\tilde{\gamma}^{a}, \tilde{\gamma}^{b}$

$$\sigma^{ab} = \frac{i}{2}[\tilde{\gamma}^{a}, \tilde{\gamma}^{b}], \qquad \eta^{ab} = \frac{1}{2}\{\tilde{\gamma}^{a}, \tilde{\gamma}^{b}\} \qquad (56)$$

Thus, the Dirac spin-1/2 fermion Lagrangian reads

$$\mathcal{L}_{\text{QEDfermions}} = \bar{\psi}\gamma^{\mu}(iD_{\mu} - eA_{\mu})\psi \equiv \bar{\psi}\gamma^{\mu}(iD_{\mu})\psi, (57)$$

where the Dirac matrices γ^{a}, γ^{b} satisfy

$$\{\gamma^{\mu}, \gamma^{\nu}\} = 2g^{\mu\nu} \qquad (58)$$

The actions of gravity and QED-fermions are given by the total action, which is recurrent

$$\Im_{TOTAL} = \Im_{G} + \Im_{\text{QED}-\text{fermions}} \qquad (59)$$

and their integrals, including the curvature tensor, are given in [24, 26].

The curvature in the second order is given as the 2-form from the notation

$$R^{cb} = d\omega^{ab} + \omega_{c}^{a} \wedge \omega^{cb} \qquad (60)$$

where $\omega^{ab} = \varpi^{ab}_{\mu} dx^{\mu}$. The closed relation between the torsion and the induced 4-Fermi interactions [26] (*which appear from the first order*) and the Einstein-Cartan theory [27] (which is our frame of reference) brings us to the study of the effects of torsion in first-order formalism, for which it is convenient to decompose the torsion tensor into irreducible components.

$$T_{\mu\nu\rho} = \frac{1}{3}(T_{\nu} g_{\mu\rho} - T_{\rho} g_{\mu\nu}) - \frac{1}{6} \in_{\mu\nu} S^{\sigma} + q_{\mu\nu\rho}, (61)$$

Then, we use the following triple product of the Dirac matrices

$$\widetilde{\gamma}^{a}\widetilde{\gamma}^{b}\widetilde{\gamma}^{c} = \eta^{ab}\widetilde{\gamma}^{c} + \eta^{bc}\widetilde{\gamma}^{a} - \eta^{ac}\widetilde{\gamma}^{b} - i \in^{abcd} \widetilde{\gamma}_{5}\widetilde{\gamma}_{d}, (62)$$

where $\widetilde{\gamma}_5 = i\widetilde{\gamma}^0\widetilde{\gamma}^1\widetilde{\gamma}^2\widetilde{\gamma}^3$

$$S_{QED-Fermions} = \frac{i}{2}\int d^4 x e(e^{\mu}_{a}(\overline{\psi}\widetilde{\gamma}^{a}\mathcal{D}_{\mu}(\varpi, A)\psi -$$

$$-\overline{\mathcal{D}_{\mu}(\varpi, A)\psi}\widetilde{\gamma}^{a}\psi) + \frac{i}{4} S_{\mu} J^{\mu}_{(A)} \qquad (63)$$

However, considering the terms $\frac{1}{24} S_{\mu} S^{\mu}, \frac{2}{3} T_{\mu} T^{\mu}, \frac{1}{2} q_{\mu\nu\rho} q^{\mu\nu\rho},$ and $\frac{i}{4} S_{\mu} J^{\mu}_{(A)},$ the variation of the combined actions $S_G + S_{QED-fermions}$, written through the *T*, *S*, and *q*-components of torsion and viewed as independent fields, yields

$$T^{\mu} = 0, \qquad \frac{1}{24\kappa} S^{\mu} - \frac{1}{8} J^{\mu}_{(A)} = 0, \qquad q_{\mu\nu\rho} = 0 \qquad (64)$$

where $J^{\mu}_{(A)} \equiv \overline{\psi}\gamma_5\gamma^{\mu}\psi$ is the *spinor axial current*. These solutions allow for the spin connection with torsion to be written as

$$\omega^{ab}_{\mu} = \varpi^{ab}_{\mu} + \frac{\kappa}{4} \in^{ab}_{cd} e^{c}_{\mu} J^{\mu}_{(A)} \qquad (65)$$

so that torsion is associated with the spinor axial current.

This treatment appears planar in some aspects in relation to possible *sources of torsion*, since J and S are linked in the expression, which is possibly of a major order, when we analyze the interactions between fermions of many generations (dispersion). In torsion phenomena, there are many aspects concerning the effects obtained by interactions between fermions and the role of neutrino production engaged at the beginning of the universe, as we can see when describing gravitational phenomena. The link between torsion and the proper fermionic nature of the beginning of the universe cannot be ignored.

In a previous paper [25], the important point was to obtain a metric constituted of the forms of Cartan-Einstein theory, with a new Cartan-Einstein theory designed to measure the "little distorsions" in the Riemannian manifold model with underlying fermionic Fock space (and Hilbert space components), suggesting fine states as observable aspects of the torsion field—this is precisely the idea that is needed.

Critics of experimental observations focus on the analysis of the problem of the behavior of neutrinos, which is unclear in the microwave model, rather than the existence of a device sufficiently fine to measure the interactions relevant in measuring torsion.

4.2. Kalb-Ramond Torsion and Axion Fields

The Dirac Lagrangian corresponding to the Majorana fields in this interaction is given as

$$\mathcal{L} = \sqrt{-g}\,(i\bar{\psi}\gamma^{a}D_{a}\psi - m\bar{\psi}\psi) \qquad (66)$$

with the derivative D_a (*gravitational covariant derivative*)

$$D_a = \left(\partial_a - \frac{i}{4}\omega_{bca}\sigma^{bc}\right) \qquad (67)$$

which includes the spin connection $\sigma^{ab} = \frac{1}{2}[\gamma^a, \gamma^b]$ and the proper gravitational connection

$$\omega_{bca}=e_{b\lambda}(\partial_a e_c^\lambda+\Gamma^\lambda_{\gamma\mu}e_c^\gamma e_a^\mu) \qquad (68)$$

then, the Lagrangian given in (16) can be rewritten as the sum of two Lagrangian components—one for the fermion and the other including the Majorana states

$$\mathcal{L}=\mathcal{L}_f+\mathcal{L}_I=\sqrt{-g}\psi[(i\gamma^a\partial_a-m)+\gamma^a\gamma^5 B_a]\psi, (69)$$

where

$$B^d=\in^{abcd}e_{b\lambda}(\partial_a e_c^\lambda+\Gamma^\lambda_{\alpha\mu}e_c^\alpha e_a^\mu) \qquad (70)$$

For the Majorana neutrinos, $\mathcal{L}_I$ above is given as

$$\mathcal{L}_I=\psi_L^\dagger\gamma^a\psi_L B_a, \qquad \mathcal{L}_I=-\psi_L^{c\dagger}\gamma^a\psi^c{}_L B_a, (71)$$

However, the fermions in a gravitational field appear in first-order formalism with the spin connection and the *veirbeins* being treated as independent and the torsion does not vanish. Indeed, the gravitational covariant derivative including spin connection is given as

$$\sigma^{ab}=\frac{i}{2}\left[\gamma^a,\gamma^b\right] \qquad (72)$$

given by (67) where the 3-form ω_{bca} is

$$\omega_{bca}=e_{b\lambda}(\partial_a e_c^\lambda+\Gamma^\lambda_{\gamma\mu}e_c^\gamma e_a^\mu) \qquad (73)$$

The existence of torsion can be verified by $\Gamma_{\mu\nu}\neq\Gamma_{\nu\mu}$ where the anti-symmetric part is the *contorsion tensor*, which contributes to

$$B^d=\in^{abcd}e_{b\lambda}(\partial_a e_c^\lambda+\Gamma^\lambda_{\alpha\mu}e_c^\alpha e_a^\mu) \qquad (74)$$

Here, the tetrads are independent of the spin connection ω_{μ}^{ab} and the field theories with Kalb-Ramond torsion and axion fields (*the Kalb-Ramond field*)[19] conform to the UV-complete models of quantum gravity and, in another way, the Majorana neutrinos masses from (three-loop) anomalous terms with axion-neutrino couplings will be the tools to develop the torsion-matter models that we want.

The microscopic UV-complete underlying theory of quantum gravity is shaped by strings, for example, the massless gravitational multiplet of (closed) strings as spin 0, scalar or dilaton; spin 2, traceless symmetric rank 2, tensor graviton; and spin 1, anti-symmetric rank 2 tensor.

The Kalb-Ramond field complies with

$$B_{\mu\nu} = -B_{\nu\mu} \qquad (75)$$

where for effective field theories (of low energy scale $E << M_s$) of "gauges", the invariants have the correspondence

$$B_{\mu\nu} \rightarrow B_{\mu\nu} + \partial_{[\mu}\theta(x)_{\nu]} \qquad (76)$$

which depends only on the field strength

$$H_{\mu\nu\rho} = \partial_{[\mu}B_{\nu\rho]} \qquad (77)$$

However, for the Bianchi identity we have

$$\partial_{[\sigma}H_{\mu\nu\rho]} = 0 \qquad (78)$$

This implies

$$d \star \mathbf{H} = 0 \qquad (79)$$

An anomaly in the cancellation between gravity vs. gauge in strings requires a redefinition of *H*, thus the Bianchi identity is now extended to

---

[19] These are string-inspired models and loop quantum gravity effective field theories.

$$\mathbf{H} = \mathbf{dB} + \frac{\alpha'}{8\kappa}(\Omega_L - \Omega_V) \qquad (80)$$

where (L) is the Lorentz force and V is the gauge in the Chern-Simons three-dimensional forms Ω. Here, $\kappa^2 = 8\pi G_N = \dfrac{8\pi}{M_P^2}$.

Thus, the differential of (64) is

$$\mathbf{dH} = \frac{\alpha'}{8\kappa}\mathrm{Tr}(\mathbf{R} \wedge \mathbf{R} - \mathbf{F} \wedge \mathbf{F}) \qquad (81)$$

As such, the role of the Kalb-Ramond **H**-fields as torsion is found in the creation of an effective gravitational action in a string with a low-energy limit, which can be expressed in terms of a generalized Riemannian curvature tensor (as desired in Cartan-Einstein theory) where the Christoffel connection includes the **H**-field torsion (in the 4-dimensional part).

$$\begin{aligned} S^{(4)} &= \int d^4x\sqrt{-g}\left(\frac{1}{2\kappa^2}R - \frac{1}{6}H_{\mu\nu\rho}H^{\mu\nu\rho}\right) \\ &= \int d^4x\sqrt{-g}\left(\frac{1}{2\kappa^2}\overline{R} - \frac{1}{3}H_{\mu\nu\rho}H^{\mu\nu\rho}\right) \end{aligned} \qquad (82)$$

where $\overline{\Gamma}^{\mu}_{\nu\rho} = \Gamma^{\mu}_{\nu\rho} + \dfrac{\kappa}{\sqrt{3}}H^{\mu}_{\nu\rho} \neq \overline{\Gamma}^{\mu}_{\rho\nu}$.

In four dimensions, we define the dual **H** as

$$-3\sqrt{2}\partial_\sigma b = \sqrt{-g}\in_{\mu\nu\rho\sigma}H^{\mu\nu\rho} \qquad (83)$$

where $b(x)$ is a pseudo-scalar or Kalb-Ramond axion. The fermions coupled to the **H**-torsion via the gravitational covariant derivative have the action (in Majorana states)

$$S_{\psi} = \frac{i}{2}\int d^4x\sqrt{-g}\,(\overline{\psi}\gamma^{\mu}\overline{\mathcal{D}}_{\mu}\psi - (\overline{\mathcal{D}}_{\mu}\overline{\psi})\gamma^{\mu}\psi), (84)$$

and we have the torsionful connection, which is a first order formalism

$$\overline{\mathcal{D}}_{\mu} = \overline{\nabla}_{\mu} - ieA_{\mu}, \qquad \varpi_{ab\mu} = \omega_{ab\mu} + K_{ab\mu} \qquad (85)$$

with the gauge field expressed by the term $-ieA_{\mu}$ and the term of contorsion given by $K_{ab\mu}$. This last one is given by

$$K_{ab\mu} = \frac{1}{2}(T_{cab} - T_{abc} - T_{bca}) \qquad (86)$$

The non-trivial contributions to B^{μ} are, as stipulated in (70), evolving as H_{cab}. As such, we amy ask whether the term of the **H**-field given in $\overline{\Gamma}^{\lambda}_{\alpha\mu}$ suggests that H is constant?

Doing exact solutions for string theory (using conformal field theories on the world-sheet) [31-33], we can give cosmological solutions in this sense.

Likewise, we have non-trivial time dependent dilatons, that is to say, *axions in an Einstein frame* (E) (the scalar curvature term in effective gravitational action has canonical normalization).

$$ds^2 = g^{E}_{\mu\nu}(x)dx^{\mu}dx^{\nu} = dt^2 - a(t)^2\delta_{ij}dx^i dx^j, (87)$$

$$a(t) = t, \quad \Phi = -\ln a(t) + \phi_0, \qquad (88)$$

$$H_{\mu\nu\rho} = e^{2\Phi}\in_{\mu\nu\rho\sigma}\partial^{\sigma}b(x), \qquad b(x) = \sqrt{2}e^{-\phi_0}\sqrt{Q^2}\frac{M_s}{\sqrt{n}}t, (89)$$

The central charge of the underlying worldsheet conformal field theory $n\in\mathbb{Z}^{+}$ is given by

$$c = 4 - 12Q^2 - \frac{6}{n+2} + c_I \qquad (90)$$

where c_I is the "internal" dimensions of the central charge in a Kac-Moody algebra level described by $n \in \mathbb{Z}^{+}$.

Thus, the conformal invariance (worldsheet) conditions for the string of order $o(\alpha')$ (*lowest non-trivial order in derivatives in the effective target-space actions*) are

$$\beta_{\mu\nu}^{g(1)} = R_{\mu\nu} + 2\nabla_{\mu}\nabla_{\nu}\phi - \frac{1}{4} H_{\mu\rho\sigma} H_{\nu}^{\rho\sigma} = 0 \qquad (91)$$

and

$$\beta_{\mu\nu}^{B(1)} = -\frac{1}{2}\nabla^{\rho} H_{\rho\mu\nu} + H_{\rho\mu\nu} = 0 \qquad (92)$$

and finally

$$\beta^{\phi(1)} = \frac{D-26}{6\alpha'} - \frac{1}{2}\nabla^{2}\phi + \partial^{\rho}\phi\partial_{\rho}\phi - \frac{1}{24} H_{\mu\nu\rho} H^{\mu\nu\rho} = 0, (93)$$

A linear dilaton solution in the string frame (or logarithmic in FRW[20] time in the Einstein-frame with conformally-flat Einstein-frame target space-time) implies exactness to all orders in α'.

4.3. H-States and CPTV

We consider the covariant torsion tensor

[20] Also known as the Friedmann-Lemaître-Robertson-Walker (FLRW) metric, is an exact solution of Einstein's field equations of general relativity; it describes a homogeneous and isotropically expanding or contracting universe that may be simply or multiply connected [34].

$$\overline{\Gamma}^{\lambda}_{\mu\nu} = \Gamma^{\lambda}_{\mu\nu} + e^{-2\Phi} H^{\lambda}_{\mu\nu} \equiv \Gamma^{\lambda}_{\mu\nu} + T^{\lambda}_{\mu\nu} \quad (94)$$

Then, considering $b(x)$ in (89), we have

$$T_{ijk} \approx \in_{ijk} \dot{b} = \in_{ijk} \sqrt{2Q^2} e^{-\phi_0} \frac{M_s}{\sqrt{n}} \quad (95)$$

the constant term of which implies that B^0 is constant. That is to say, we can approximately calculate B^0 as [33, 35-38]

$$B^0 \approx \sqrt{2Q^2} e^{-\phi_0} \frac{M_s}{\sqrt{n}} (= GeV) > 0 \quad (96)$$

Lepton asymmetry is observed as in previous cases, e.g. for neutrinos

$$\Delta L(T - T_d) = \frac{\Delta n_\nu}{s} \approx \frac{B_0}{T_d} \quad (97)$$

but the dispersion relations of the neutrinos are different from those of anti-neutrinos in such geometries!

If we consider the neutrino mass relation by the discriminant $(p_a \pm B_a)^2 = m^2$ (where $\pm$ refers to chiral fields (*here neutrino/anti-neutrino*)), the CPTV dispersion relations given by

$$E = \sqrt{p^2 + m^2} + B_0, \qquad E = \sqrt{p^2 + m^2} - B_0 \quad (98)$$

show that the (bare) masses are equal between the particle/anti-particle sectors. In fact, the abundance of neutrinos in the early universe implies a difference to anti-neutrinos if B^0 is non-trivial, which is always already in thermal equilibrium.

As such, was it necessary to conserve sphalerons in CPTV baryogenesis through B-L-violation? That is to say, was there any need for enhanced *CP*-violation in the early universe?

The answer is very hard to give, but we can say that CPTV implies neutrino/anti-neutrino mixing and oscillations [44, 45]. This mixture of oscillations and neutrinos/anti-neutrinos can be measured through spinors. However, we need to establish the front of the "waves" at this scale.

We consider the Lagrangian of the Majorana states

$$\mathcal{L} = \det(e)\Psi\left(\frac{i}{2}\gamma^a \overleftrightarrow{\partial}_a - m + \gamma^a\gamma^5 B_a\right)\Psi \qquad (99)$$

where $B^d = \in^{abcd} \omega_{bca}$. Then for neutrinos ν and ν^c we can consider

$$E_\nu = \sqrt{(p-B)^2 + m^2} + B_0$$
$$E_{\nu^c} = \sqrt{(p+B)^2 + m^2} - B_0 \qquad (100)$$

Assuming a Majorana neutrino in a Weyl representation (*lepton number violation being unavoidable*), we have

$$\Psi = \begin{pmatrix} \psi_L^c \\ \psi_L \end{pmatrix}, \qquad \mathcal{D}_\mu = (\partial_0, \partial_i + \gamma^5 B_i) \qquad (101)$$

Thus, CP-violating neutrino/anti-neutrino mixing and oscillations lead us to

$$(-g)^{-1/2}\mathcal{L} = (\psi^{c\dagger}, \psi^\dagger)\frac{i}{2}\gamma^0\gamma^\mu \overleftrightarrow{\mathcal{D}}_\mu \begin{pmatrix} \psi^C \\ \psi \end{pmatrix} -$$
$$(\psi^{c\dagger}, \psi^\dagger)\begin{pmatrix} -B_0 & -m \\ -m & B_0 \end{pmatrix} \qquad (102)$$

with mass eigenstates ν_1 and ν_2

$$|\nu_1\rangle = \frac{1}{N}\left\{\left(B_0 + \sqrt{B_0^2 + m^2}\right)|\psi^c\rangle + m|\psi\rangle\right\} \qquad (103)$$

and

$$|\nu_2\rangle = \frac{1}{N}\left\{-m|\psi^c\rangle + \left(B_0 + \sqrt{B_0^2 + m^2}\right)|\psi\rangle\right\} \tag{104}$$

with $m_{1,2} = \mp\sqrt{B_0^2 + m^2}$. The eigenstates (103) and (104), through a planar rotational transformation, take the form

$$|\nu_1\rangle = \cos\theta|\psi^c\rangle + \sin\theta|\psi\rangle \tag{105}$$

and

$$|\nu_2\rangle = -\sin\theta|\psi^c\rangle + \cos\theta|\psi\rangle \tag{106}$$

with a shift in neutrino CPTV mass

$$\tan\theta = \frac{m}{B_0 + \sqrt{B_0^2 + m^2}} \tag{107}$$

We have the neutrino/antineutrino mixing and oscillations given by the formalism

$$|\psi^c\rangle = \cos\theta|\nu_1\rangle - \sin\theta|\nu_2\rangle \tag{108}$$

$$|\psi\rangle = \sin\theta|\nu_1\rangle + \cos\theta|\nu_2\rangle \tag{109}$$

$$\mathcal{P}(t) = \sin^2 2\theta \sin^2 \delta(t) \tag{110}$$

$$Oscillations = \frac{m^2}{B_0^2 + m^2}\sin^2\left\{(B_0 - |\vec{B}|)t\right\}, \tag{111}$$

and, for this scenario, we can obtain the following conjecture.

Conjecture 4.3.1. [33]. The phase transitions in the early string universe may have induced $B^0 = 0$ and $T = T_d$.

We know that if $T > T_d$ and $B^0 \neq 0$, then we have Majorana neutrino/anti-neutrino oscillations (*of the Pontecorvo type*). If we consider the 4-component chiral Majorana spinor Ψ, given by the Weyl representation in (101) and with Lagrangian (102), we have the following mass eigenstates with their eigenvalues

$$\left|\chi_1\right\rangle = \mathcal{N}^{-1}\left\{\left(B_0+\sqrt{B_0^2+m^2}\right)\left|\psi^c\right\rangle + m\left|\psi\right\rangle\right\} \qquad (112)$$

and

$$\left|\chi_2\right\rangle = \mathcal{N}^{-1}\left\{-m\left|\psi^c\right\rangle + \left(B_0+\sqrt{B_0^2+m^2}\right)\left|\psi\right\rangle\right\} \qquad (113)$$

where $m_{1,2} = \mp\sqrt{B_0^2+m^2}$ and

$$\mathcal{N} = \left[2(B_0^2+m^2+B_0\sqrt{B_0^2+m^2})\right]^{1/2} \qquad (114)$$

Then, the new neutrino/anti-neutrino mixing and oscillations are defined by the transformation

$$\nu \equiv \begin{pmatrix} \chi_1 \\ \chi_2 \end{pmatrix} = \begin{pmatrix} \cos\theta & \sin\theta \\ -\sin\theta & \cos\theta \end{pmatrix} \begin{pmatrix} \psi^c \\ \psi \end{pmatrix} \qquad (115)$$

Considering (107), we can define the following number operators

$$N_{\chi_1} = <\chi_1^\dagger \chi_1 :>= \cos^2\theta <:\psi^{c\dagger}\psi^c:>$$
$$+\sin^2\theta <:\psi^\dagger\psi:> \qquad (116)$$

and

$$N_{\chi_2} = <\chi_2^\dagger \chi_2 :>= \sin^2\theta <:\psi^{c\dagger}\psi^c:>$$
$$+\cos^2\theta <:\psi^\dagger\psi:> \qquad (117)$$

with their difference given by

$$N_{\chi_1} - N_{\chi_2} = \cos 2\theta(<n_{\psi^c}> - <n_\psi>) \neq 0 \qquad (118)$$

for $\theta \neq \pi/4$, which is to say $B^0 \neq 0$.

We can then calculate the probability of the oscillation.

This provides the first evidence of torsion through **H**-states (or **H**-torsion), having Majorana neutrino/anti-neutrino oscillations (see fig. 22 A), B)).

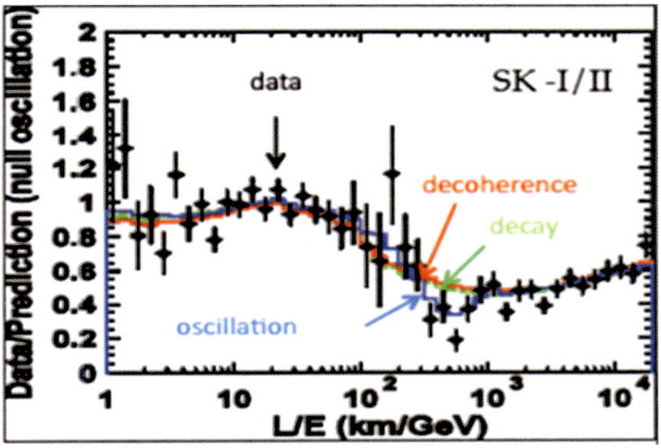

A)

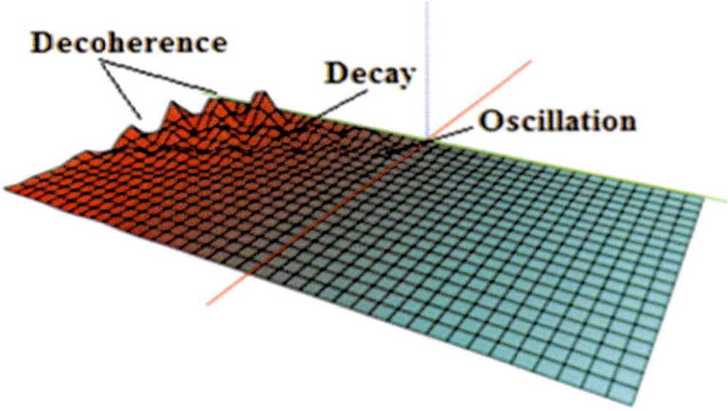

B)

Fig. 22. 2-D model of neutrino/anti-neutrino oscillations, decay, and de-coherence. This 2-D surface is considered in [27] by: A) data prediction on observational fact; B) geometrical 2-D flat model considering solutions of the Bessel functions $Z = exp(-0.5(x+1))BesselY(0.5x, 6y)sin(4x)BesselJ(0.3y,20)$. The oscillations in de Sitter space maybe happen after of de-coherence and decay due the homogeneous space-time structure of sterile neutrinos and light neutrinos in dark matter.

4.4. Field Torsion Mechanism to Create Matter and Corresponding Dirac Equations of Neutrinos and Matter/anti-Matter Asymmetries

4.4.1. General Discussion

Observing the predominance of matter over anti-matter in the universe and the creation of matter through mechanisms that rely on the smallness of mass of active neutrino species, we can define the material universe as space-time at the quantum level created through fragments by the breaking of heavy particles like baryons. These aspects display asymmetries that may explain the proliferation of leptons in the polarizable constitution of neutrinos, which conform to quantum torsion in terms of Majorana neutrino masses and from which dark matter is formed as an ulterior product.

One reason for this conclusion is the *KR*-background torsion, which yields matter/anti-matter asymmetry in the frame of gauge fields. That is to say, we can use the bosonic *KR*-torsion from string theory to generate the matter/anti-matter asymmetry in the string model of the universe.

These boson fields rely on the use of pseudo-scalar fields, as a new version of scalar field theory with infinite extensions of the Standard Model (SM), being responsible for both right-handed Majorana neutrino masses and the homogeneity of the Dirac equations, and thus we do not need to make extra adjustments in response to any *CP*-violation. As such, we have two Dirac equations corresponding to two Majorana neutrino masses and their relationship shapes the quantum torsion as we wish.

4.4.2. From Einstein-Cartan Formalism to Fermions in the Gravitational Field

In Einstein's theories (of general and special relativity) torsion has been absent of constraint as a field observable and we consider it an effect that depends on the interactions of matter/anti-matter, which is not suggested when treating torsion as a dynamic field. As such, we do not consider torsion to be independent of the vierbeins (or the Cartan connection [23]).

Fermions appear in the gravitational field when the spin connection is formulated in first-order formalism; the veirbeins are treated as being independent and the torsion does not vanish.

The relations between gravity and torsion are linked by gravity and asymmetries established in terms of *gravity with torsion* whose connection is

$$\omega_{\mu}^{ab} = \overline{\omega}_{\mu}^{ab} + K_{\mu}^{ab}$$

where the connection $\overline{\omega}_{\mu}^{ab}$ is given by

$$\overline{\omega}_{\mu}^{ab} = e_{\nu}^{a}\partial_{\mu}e^{\nu b} + e_{\nu}^{a}e^{\sigma b}\Gamma_{\sigma\mu}^{\nu} = e_{\nu}^{a}e_{\mu}^{\nu b}$$

The other connection is given for K_{μ}^{ab}

$$K_{\mu}^{ab} = e_{\nu}^{a}e_{\rho}^{b}K_{\mu}^{\nu\rho}$$

such that $K_{\mu}^{\nu\rho} = -K_{\mu}^{\rho\nu}$, where K_{μ}^{ab} is the *contorsion tensor*.

Thus, torsion $T_{\nu\rho}^{\mu}$ can be written according to the covariant contorsion tensor as

$$K_{\rho\mu}^{\nu} = \frac{1}{2}(T_{\rho\mu}^{\nu} - T_{\rho}{}^{\nu}{}_{\mu} - T_{\mu}{}^{\nu}{}_{\rho}) \tag{119}$$

The curvature tensor in first-order torsionful formalism comes to

$$R_{\mu\nu}^{ab} = 2\partial_{[\mu,}\omega_{\nu]}^{ab} + 2\omega_{c[\mu}^{a}\omega_{\nu]}^{cb} \tag{120}$$

For the other side, the covariant derivative of the fermions is

$$D_\mu \psi = \left(\partial_\mu - \frac{i}{4} \omega_\mu^{ab} \sigma_{ab} \right) \psi \qquad (121)$$

with the tensors from the Dirac matrices $\widetilde{\gamma}^a, \widetilde{\gamma}^b$ of

$$\sigma^{ab} = \frac{i}{2}[\widetilde{\gamma}^a, \widetilde{\gamma}^b], \qquad \eta^{ab} = \frac{1}{2}\{\widetilde{\gamma}^a, \widetilde{\gamma}^b\} \qquad (122)$$

The Dirac spin-1/2 fermion Lagrangian thus reads

$$\mathcal{L}_{\text{QED - fermions}} = \overline{\psi} \gamma^\mu (iD_\mu - eA_\mu)\psi \equiv \overline{\psi} \gamma^\mu (iD_\mu)\psi, (123)$$

where the Dirac matrices γ^a, γ^b satisfy

$$\{\gamma^\mu, \gamma^\nu\} = 2g^{\mu\nu} \qquad (124)$$

The action of gravity and the QED-fermions is given by the total action of

$$\Im_{TOTAL} = \Im_G + \Im_{\text{QED-fermions}} \qquad (125)$$

and their integrals, including the curvature tensor, are given in [14, 24-26].

Then, curvature in the second order is given in the 2-form notation

$$R^{ab} = d\omega^{ab} + \omega^a_c \wedge \omega^{cb} \qquad (126)$$

where $\omega^{ab} = \varpi_\mu^{ab} dx^\mu$.

Thus, the closed relation between the torsion, the induced 4-Fermi interactions [26], and Einstein-Cartan theory [27] comes via the study of the effects of torsion in first-order formalism and it is convenient to decompose the torsion tensor into irreducible components according to their transformation under the local Lorentz group of transformations

$$T_{\mu\nu\rho} = \frac{1}{3}(T_\nu g_{\mu\rho} - T_\rho g_{\mu\nu}) - \frac{1}{6} \in_{\mu\nu} S^\sigma + q_{\mu\nu\rho}, (127)$$

where $T_\mu \equiv T^\nu_{\mu\nu}$ is the torsion trace vector; $S_\mu \equiv \in_{\mu\nu\rho\sigma} T^{\nu\mu\sigma}$ is the pseudo-trace axial vector; and $q^\nu_{\rho\nu} = 0 = \in^{\sigma\mu\nu\rho} q_{\mu\nu\rho}$ is the anti-symmetric tensor.

The gravitational action written as the precursor tensors takes the form

$$S_G = \frac{1}{2k}\int d^4x e(e^\mu_a e^\mu_b R^{ab}_{\mu\nu}(\varpi) + \frac{1}{24}S_\mu S^\mu - \frac{2}{3}T_\mu T^\mu + \\ + \frac{1}{2}q_{\mu\nu\rho}q^{\mu\nu\rho} \qquad (128)$$

We use the following triple product of the Dirac matrices

$$\tilde{\gamma}^a\tilde{\gamma}^b\tilde{\gamma}^c = \eta^{ab}\tilde{\gamma}^c + \eta^{bc}\tilde{\gamma}^a - \eta^{ac}\tilde{\gamma}^b - i\in^{abcd}\tilde{\gamma}_5\tilde{\gamma}_d, (129)$$

where $\tilde{\gamma}_5 = i\tilde{\gamma}^0\tilde{\gamma}^1\tilde{\gamma}^2\tilde{\gamma}^3$. Thus, we obtain

$$S_{QED-Fermions} = \frac{i}{2}\int d^4x e(e^\mu_a(\overline{\psi}\tilde{\gamma}^a\mathcal{D}_\mu(\varpi, A)\psi - \\ -\overline{\mathcal{D}_\mu(\varpi, A)\psi}\tilde{\gamma}^a\psi) + \frac{i}{4}S_\mu J^\mu_{(A)} \qquad (130)$$

Considering the terms $\frac{1}{24}S_\mu S^\mu, \frac{2}{3}T_\mu T^\mu, \frac{1}{2}q_{\mu\nu\rho}q^{\mu\nu\rho}$, and $\frac{i}{4}S_\mu J^\mu_{(A)}$, the variation of the combined actions $S_G + S_{QED-fermions}$, written in terms of the T, S, and q components of torsion and viewed as independent fields, yields

$$T^\mu = 0, \qquad \frac{1}{24\kappa}S^\mu - \frac{1}{8}J^\mu_{(A)} = 0, \qquad q_{\mu\nu\rho} = 0 \qquad (131)$$

where $J^\mu_{(A)} \equiv \overline{\psi}\gamma_5\gamma^\mu\psi$ is the spinor axial current. These solutions allow for the spin connection with torsion to be written as

$$\omega^{ab}_\mu = \varpi^{ab}_\mu + \frac{\kappa}{4}\in^{ab}_{cd} e^c_\mu J^\mu_{(A)} \qquad (132)$$

Thus, the torsion is associated with the spinor axial current.

4.4.3. The Total Action in Cartan-Einstein Theory

We consider the total action involving the integral term of torsion as

$$S_{TOTAL} = \frac{1}{2\kappa}\int d^4 x e e_a^\mu e_b^\nu R_{\mu\nu}^{ab}(\varpi) +$$
$$\frac{i}{2}\int d^4 x e (e_a^\mu (\overline{\psi}\widetilde{\gamma}^a \mathcal{D}_\mu \varpi, A)\psi - \overline{\mathcal{D}_\mu(\varpi, A)\psi}\widetilde{\gamma}^a\psi)$$
$$+\int d^4 x e \frac{3}{16}\kappa J_{(A)}^\mu J_{(A)\mu} \quad (133)$$

where $\kappa = 8\pi G_N$ is the gravitational constant and $\kappa J_{(A)}^\mu J_{(A)\mu}$ is the fermion interactions induced by the quantum torsion.

Why is an understanding of torsion required for studying the mechanisms of matter/anti-matter in gravity and neutrinos/anti-neutrinos in the production of matter to develop gravity? What species of neutrinos are relevant to this?

To derive a Dirac model that explains gravity generated by the action of neutrinos and other fermions with torsion, it is necessary to contemplate the following short summary of torsion connections. First, we can consider the connection given by (132) with $J_{(A)}^\mu \equiv \overline{\psi}\gamma_5\gamma^\mu\psi$. In the case of $N = 1$ SUGRA, the connection is

$$\omega_\mu^{ab} = \omega_\mu^{ab}(e,\psi) \equiv \omega_\mu^{ab}(e) + \kappa_\mu^{ab}(\psi) \quad (134)$$

where

$$\kappa_{\mu\alpha\beta}(\psi) = \frac{1}{4}[\overline{\psi}_\alpha\gamma_\mu\psi_\beta + \overline{\psi}_\mu\gamma_\alpha\psi_\beta - \overline{\psi}_\mu\gamma_\beta\psi_\alpha], \quad (135)$$

and the spinor axial current is

$$J^\mu \equiv \frac{1}{2} e^{\mu\nu\alpha\beta}\, \overline{\psi}_\nu\gamma_\alpha\psi_\beta \quad (136)$$

For other side, in the case of $N = 2$ SUGRA the connection is the same as that in (134), with the difference that $\kappa_{\mu\alpha\beta}(\psi)$ is given by

$$\kappa_{\mu\alpha\beta}(\psi) = \frac{1}{4}[\overline{\psi}_{\nu}^{l}\gamma_{\mu}\psi_{l\beta} + \overline{\psi}_{\mu}^{l}\gamma_{\alpha}\psi_{l\beta} - \overline{\psi}_{\mu}^{l}\gamma_{\beta}\psi_{l\alpha} + c.c.], (137)$$

with the spinor axial current

$$J^{\mu} \equiv e^{\mu\nu\alpha\beta}\overline{\psi}_{\nu}^{l}\gamma_{\alpha}\psi_{l\beta} \qquad (138)$$

For the last one, in the case of $N = 14$ SUGRA the connection is

$$\omega_{\mu ab} = \omega_{\mu ab}(e,\psi,\Lambda) \equiv \omega_{\mu ab}(e) + \kappa_{\mu ab}(\psi) \qquad (139)$$

where

$$\kappa_{\mu\alpha\beta} = \kappa_{\mu\alpha\beta}(\psi) + \kappa_{\mu\alpha\beta}(\Lambda) \qquad (140)$$

The term given by $\kappa_{\mu\alpha\beta}(\psi)$ is the same as (23), while the term $\kappa_{\mu\alpha\beta}(\Lambda)$ is given by

$$\kappa_{\mu\alpha\beta}(\Lambda) = -\frac{1}{4e}\in_{\mu\alpha\beta\sigma}\overline{\Lambda}_{I}\gamma^{\sigma}\Lambda^{I} \qquad (141)$$

where $J^{\mu}(\Lambda) = e\overline{\Lambda}_{I}\gamma^{\mu}\Lambda^{I}$.

However, some ambiguities exist in Einstein-Cartan theory in relation to the Immirzi parameter. For example, actions can be modified by total derivatives without affecting the on-shell physics (that is to say, for equations of motion), but such terms may play a role in quantum physics. On the other hand, in the case of pure gravity without torsion, one can add the following identically zero term due to the Bianchi identities of the symmetry properties of the Riemann tensor

$$\in^{\mu\nu\rho\sigma} R_{\mu\nu\sigma\rho}(\varpi)=0 \qquad (142)$$

which is no longer true in the presence of torsion. For example, where the tensor $\widetilde{R}^{ab}_{\mu\nu} = \in^{ab}_{cd} R^{cd}_{\mu\nu}$ is the dual of the curvature tensor, the action

$$S_{Holst} = -\frac{\beta}{4\kappa}\int d^4 x e e^{\mu}_{a} e^{\mu}_{b} \in^{ab}_{cd} R^{cd}_{\mu\nu} \qquad (143)$$

is nontrivial in terms of torsionful geometries. However, in the case of torsion, the equations of movement imply

$$T^{\mu} = \frac{3\kappa}{4}\frac{\beta}{\beta^2+1} J^{\mu}_{(A)} \qquad (144)$$

$$S^{\mu} = \frac{3\kappa}{\beta^2+1} J^{\mu}_{(A)} \qquad (145)$$

$$q_{\mu\nu\rho} = 0 \qquad (146)$$

which is clearly inconsistent due to the equality between the vector T^{μ} and the pseudo-vector $J^{\mu}_{(A)}$.

The Holst term needs modification by the addition of fermions to become a total derivative [22]. The fermion-piece that will be added to the Holst action is

$$S_{Holst-fermion} = +\frac{\alpha}{2}\int d^4 x e \times$$
$$\times \left(\overline{\psi}\gamma^{\mu}\gamma_5 \mathcal{D}_{\mu}(\omega)\psi + \overline{\mathcal{D}_{\mu}(\omega)}\gamma^{\mu}\gamma_5\psi\right), \alpha = cont \qquad (147)$$

Together with the Dirac kinetic term, the fermion action reads

$$S_{Dirac-Holst-fermion} = \frac{i}{2}\int d^4xe \times$$
$$\times \left(\overline{\psi}\gamma^{\mu}(1-i\alpha\gamma_5)\mathcal{D}_{\mu}(\omega)\psi + \overline{\mathcal{D}_{\mu}(\omega)}\gamma^{\mu}(1-i\alpha\gamma_5\psi)\right) \tag{148}$$

If $\alpha = \beta$, then the *Holst total gravity-fermion* term becomes a total derivative, which is to say

$$S_{Holst-total} = -\frac{\beta}{2}\int d^4x\left[I_{NY} + \partial_{\mu}J^{\mu}_{(A)}\right] \tag{149}$$

This establishes an equivalent to Einstein-Cartan theory independent of the Immirzi parameter β.

In (149), I_{NY} is the *Nieh-Yan topological invariant density*, creating a unique Lorentz invariant torsion structure that has the form

$$I_{NY} \equiv e^{\mu\nu\rho\sigma}\left(T^{a}_{\mu\nu}T_{\rho\sigma a} - \frac{1}{2}e^{a}_{\mu}e^{b}_{\nu}R_{\rho\sigma ab}(\omega)\right) \tag{150}$$

In simplified terms, this can be written

$$I_{NY} \equiv e^{\mu\nu\rho\sigma}\partial_{\mu}T_{\mu\nu\sigma} \tag{151}$$

and we can draw up the following summary of (super)Holst modifications (see table 2 below).

TABLE 2: (Super)Holst Modifications)

Spinor Axial Current	**Interaction**	**Action Integral**
$J^{\mu}_{(A)} \equiv \psi\gamma_5\gamma^{\mu}\psi$	**Dirac spin ½ fermions**	$S_{Holst-total} = -\frac{i\beta}{4}\int d^4x\partial_{\mu}J^{\mu}_{(A)}$ $= \frac{i\beta}{6}\int d^4x \in^{\mu\nu\rho\sigma} \partial_{\mu}T_{\nu\rho\sigma}(\psi)$
$J^{\mu}(\psi) = \frac{1}{2}\in^{\mu\nu\alpha\beta}$ $\times\psi_{\nu}\gamma_{\alpha}\psi_{\beta}$	N=1 SUGRA	$S_{SHolst1}(\omega(e,\psi)) = -\frac{i\eta}{4}\int d^4x\partial_{\mu}J^{\mu}_{(A)}$ $= \frac{i\eta}{2}\int d^4x \in^{\mu\nu\alpha\beta} \partial_{\mu}T_{\nu\alpha\beta}(\psi)$
$J^{\mu}(\psi) = \frac{1}{2}\in^{\mu\nu\alpha\beta}$ $\times\overline{\psi}^{l}_{\nu}\gamma_{\alpha}\psi_{l\beta}$	N=2 SUGRA	$S_{SHolst2}(\omega(e,\psi)) =$ $-\frac{i\eta}{4}\int d^4x\partial_{\mu}J^{\mu}(\psi)$
$J^{\mu}(\psi) = \frac{1}{2}\in^{\mu\nu\alpha\beta}$ $\times\overline{\psi}^{l}_{\nu}\gamma_{\alpha}\psi_{l\beta}$, $J^{\mu}(\Lambda) = e\overline{\Lambda}_{I}\gamma^{\mu}\Lambda^{I}$	$N=4\,SU(4)$ SUGRA	$S_{SHolst4}(\omega(e,\psi,\Lambda)) =$ $-\frac{i\eta}{4}\int d^4x\partial_{\mu}[J^{\mu}(\psi) - J^{\mu}(\Lambda)]$

4.4.4. Scalar Fields in the Early Universe

Considering a different line of thought about bosonic torsion with string models of extended particle physics corresponding to a pseudo-scalar degree of freedom (possible string types inspired by Kalb-Ramond called axions [48]), we could establish models of extended particles physics based on the idea of field axions and their effects in space-time, such as inflatons, dilatons, axions, and moduli stacks in relation to moduli spaces $\mathfrak{M}_{H}(^{L}G, C)$ and other physics. Thus, we concentrate on the string-inspired models of torsion or those related to torsionful geometries, that is to say, torsionful geometries of the early universe due to a Kalb-Ramond pseudo-scalar or "*KR*-axion".

In this formalism, in more concrete terms, we concentrate on the following aspects:

1) Matter/anti-matter asymmetry in the universe or the leptogenesis/baryogenesis of the universe (see fig. 1). Here, right-handed neutrinos are not needed for [30] extra *CP*-violation[21].

2) Determination of *Majorana right-handed Neutrino masses* from *KR*-axions (quantum torsion) beyond the seesaw through anomalies, or other observable aspects, both theoretical and experimental. This concerns, for example, aspects of matter/anti-matter asymmetry in thermal equilibrium in the early universe due to the background [32], or the appearance of anomalous Majorana neutrino mass generation due to quantum *KR*-torsion [32, 33].

We present some general considerations on *CP*-violation. Firstly, in the Standard Model, the *CP*-violation is not sufficient to produce the baryon/anti-baryon asymmetry we observe (it is necessary to assume that, in the early universe, we have $\not{B}-, C-, \varepsilon CP-$ violations, which are known as Sakharov conditions [35]). Secondly, there are several ideas about how to go beyond the SM (that is to say, *GUT*-models, supersymmetry, and extra dimensional models, etc). Thirdly, use of massive ν offers the simplest extension of the SM. Fourthly, the use of right-handed massive ν may offer an extension of the SM with extra *CP*-violation and thus help describe the

[21] In particle physics, *CP*-violation (charge parity) is a violation of the postulated *CP*-symmetry (or charge conjugation parity symmetry): the combination of *C*-symmetry (charge conjugation symmetry) and *P*-symmetry (parity symmetry). *CP*-symmetry states that the laws of physics should be the same if a particle is interchanged with its antiparticle (*C*-symmetry) and then its spatial coordinates are inverted ("mirror" or *P*-symmetry). It plays an important role both in attempts of cosmology to explain the dominance of matter over anti-matter in the present universe and in the study of weak interactions in particle physics.

origin of the universe's matter-antimatter asymmetry due to neutrino masses, which may be dark matter (see fig. 23).

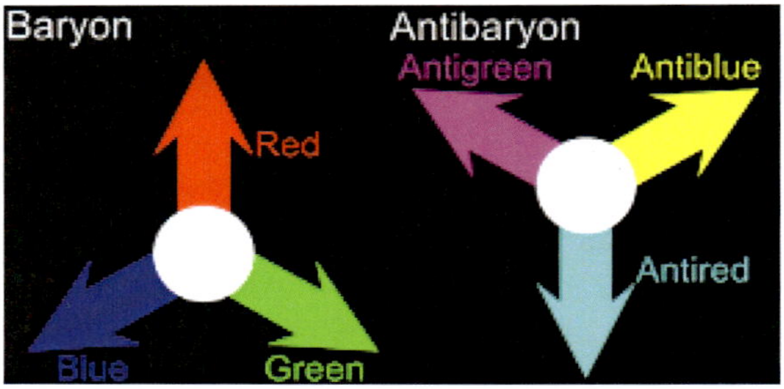

Fig. 23. Baryogenesis with mirror symmetry.

In the SM, the structures with the lowest *CP*-violation can be written according to their down quark

$$d_{CP} = \sin(\theta_{12})\sin(\theta_{23})\sin(\theta_{13})\sin\delta_{CP} \bullet$$
$$\bullet (m_t^2 - m_c^2)(m_t^2 - m_u^2)(m_c^2 - m_u^2)(m_b^2 - m_s^2) \bullet$$
$$(m_b^2 - m_d^2)(m_s^2 - m_d^2), \qquad (152)$$

where δ_{CP} is the Kobayashi-Maskawa-violating phase. Using the Shaposhnikov parameter, given by the trace of the muon product with their squares corresponding to up and down quarks, we have

$$D = \mathrm{Im}\, Tr[\mathcal{M}_u^2 \mathcal{M}_d^2 \mathcal{M}_u \mathcal{M}_d], \qquad (153)$$

The Kobayashi-Maskawa-violating phase δ_{KM}^{CP} takes the form

$$\delta_{KM}^{CP} \approx \frac{D}{T^{12}} \approx 10^{-20} << \frac{n_B - \overline{n}_B}{n_B + \overline{n}_B} \approx \frac{n_B - \overline{n}_B}{s}$$
$$= (8.4 - 8.9) \times 10^{-11} \qquad (154)$$

where $T \cong T_{sph}$ is the temperature when using hadronic gas to interpolate high and low temperatures in response to the challenge presented by the production temperature of sterile neutrinos (see fig. 83) and the quotient $\frac{n_B - \bar{n}_B}{n_B + \bar{n}_B}$, which is the baryonic asymmetry in the QCD epoch[22].

The temperature of the hadronic resonances is established in the interval $T_{sph}(m_H) \in [130 - 190]GeV$, which cannot be the source of baryon asymmetry in the universe [112].

A

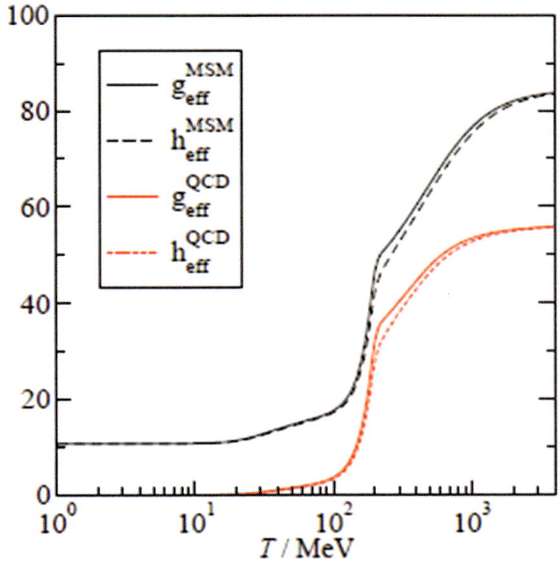

[22] If $T >> \Lambda_{QCD}$ quarks are used. For the other case, when $T << \Lambda_{QCD}$ hadrons are used. As a conservative upper bound on the hadronic contribution, we put $N_c = 0$ in the phenomenological mean value when $N_c \rightarrow N_c \frac{h_{eff}^{QCD}(T)}{58}$.

B

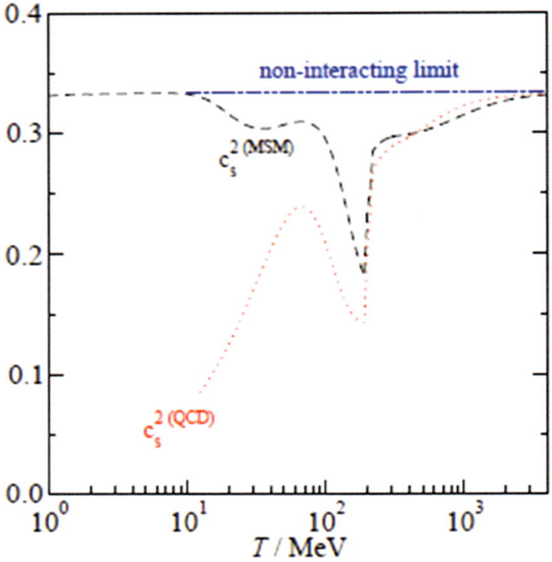

Fig. 24. Here we describe the methods of resonances of hadronic gas at low temperatures. A) The most advanced method (up to the resumption of loop weak-coupling). This results in high temperatures. B) Interpolation at intermediate temperatures.

We return to the previous remark about how right-handed massive ν may allow us to extend the *SM* with extra *CP*-violations and thus assist us in understanding the origin of the universe's matter-antimatter asymmetry due to neutrino masses, which may be dark matter.

For the SM with an extension by *N*, the extra right-handed neutrinos have Lagrangian or νMSM as

$$L = L_{SM} + \overline{N}_I i\partial_\mu \gamma^\mu - F_{\alpha I}\overline{L}_\alpha N_I \tilde{\phi} - \frac{M_I}{2}\overline{N}_I^c N_I + h.c. \qquad (155)$$

where $F_{\alpha I}$ is the Yukawa coupling matrix given in [113, 114]

$$F = \widetilde{K}_L f_d \widetilde{K}_R, \tag{156}$$

Then, light neutrino masses, through the seesaw mechanism (neutrino oscillations), can be described by

$$m_\nu = -M^D \frac{1}{M_I}[M^D]^T \tag{157}$$

where $M_D = F_{\alpha I}\upsilon$ with $\upsilon = \langle\phi\rangle \approx 175\,GeV$ and $M_D << M_I$.

In experiments using hadron gas resonances, the Yukawa coupling matrix is proved for the cases $N = 2$ or $N = 3$, in terms of the constraints (compiled or oscillation data) on light sterile neutrinos [20, 22]. The case of $N = 1$ is excluded.

A model with 2 or 3 singlet fermions works well in reproducing baryon asymmetry and is consistent with the experimental data on neutrino oscillations.

The model with $N = 3$ also works well and, in fact, it allows one of the Majorana fermions to almost decouple from the rest of the SM fields, providing candidates for light (*keV* region of mass) sterile neutrino dark matter.

Thus, as detailed in the following step, we can conclude:

- The extension with N and extra right-handed neutrinos is perturbatively consistent (stable).
- The Standard Model Higgs sector is presented.
- When right-handed neutrinos are present, it appears that we require:
  - *Three right handed-neutrinos;*
  - *Two that are degenerate in mass (heavier than GeV);*
  - *One light (in keV region) as a candidate for dark matter;*
  - *CPT is assumed with the exact symmetry of the early universe*
  - *And CP-violation is enhanced due to degeneracy.*

Furthermore, in reference to the model with $N = 3$, we may add that it also provides a consistent model for leptogenesis and, through the sphaleron process, baryogenesis.

Here, the last point may not be sufficient to reproduce baryon asymmetry, since the mechanism for baryon asymmetry through coherent oscillations of degeneration in mass neutrinos is rather delicate!

For the thermal properties, we observe that $M_I < M_W$ (*electroweak scale*), that is to say, $M_I = O(1)GeV$. This implies that we keep the light neutrino masses in right order. Thus, the Yukawa couplings must be

$$F_{\alpha I} \approx \frac{\sqrt{m_{\text{atm}} M_I}}{\upsilon} \approx 4 \times 10^{-8} \qquad (158)$$

and thus baryogenesis through coherent oscillations of right-handed singlet fermions must be considered [38, 49, 52-59].

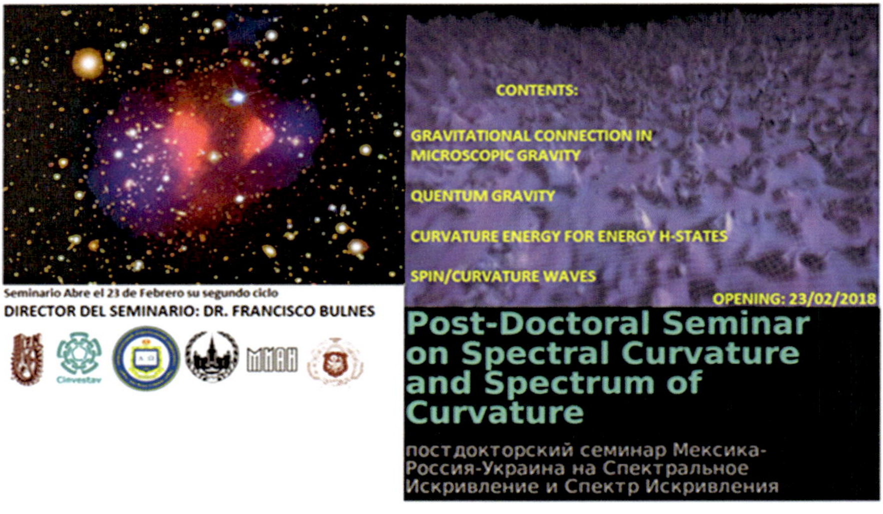

Fig. 25. Post-Doctoral Seminar on Spectral Curvature and Spectrum of Curvature, second edition, year 2018-2019. Director: Dr. Francisco Bulnes [50].

4.5. Torsion and the Corresponding Dirac Equations of Neutrinos and Matter/anti-Matter Asymmetries: Beyond Neutrinos

4.5.1. Total Action

We consider the quantum (totally anti-symmetric) torsion S^{μ}. Its action is given by

$$\int DSDBb \exp\left[i\int \frac{3}{4\kappa^2}\mathbf{S}\wedge{}^*\mathbf{S} - \frac{3}{4}\mathbf{S}\wedge{}^*\mathbf{J}^5 + \left(\frac{3}{2\kappa^2}\right)^{1/2} bd{}^*\mathbf{S}\right]$$
$$= \int Db \exp\left[-i\int \frac{1}{2}\mathbf{d}b\wedge{}^*\mathbf{d}b + \frac{1}{f_b}\mathbf{d}b\wedge{}^*\mathbf{J}^5 + \frac{1}{2f_b^2}\mathbf{J}^5\wedge\mathbf{J}^5\right] \qquad (159)$$

where the term $\frac{1}{2f_b^2}\mathbf{J}^5\wedge\mathbf{J}^5$ [23] represents the quantum-torsion-induced four-fermi interactions, not forgetting that

$$\mathbf{J}^{5\mu} = \overline{\psi}\gamma^5\gamma^{\mu}\psi \qquad (160\ )$$

and the multiplier field $\Phi(x)$ is given by

$$\Phi(x) \equiv (3/\kappa^2)^{1/2}\, b(x) \qquad (161\ )$$

even in the absence of any background field b.

For the other side, the Dirac Lagrangian comes out as

$$\mathcal{L} = \frac{i}{2}\mathfrak{e}e_a^{\mu}(\overline{\psi}\gamma^a\psi_{;\mu} - \overline{\psi}_{;\mu}\gamma^a\psi) - m\mathfrak{e}\overline{\psi}\psi \qquad (162)$$

---

[23] Here, $f_b = (3\kappa^2/8)^{-1/2} = \frac{M_P}{\sqrt{3\pi}}$.

where $\mathfrak{e} = \det(e^a_\mu)$. The variation with respect to torsion yields the connection between torsion and axial currents. Then, substitution in the fermion equations yields the higher order Hehl-Datta equation[24]

$$ie^\mu_a \gamma^a \psi_{;\mu} = m\psi - \frac{3\kappa}{8}(\overline{\psi}\gamma^5\gamma_a\psi)\gamma^5\gamma^a\psi \qquad (163)$$

obtained from the effective Lagrangian

$$\mathcal{L}_e = \frac{i}{2}\mathfrak{e}e^\mu_a(\overline{\psi}\gamma^a\psi_{;\mu} - \overline{\psi}_{;\mu}\gamma^a\psi) - m\mathfrak{e}\overline{\psi}\psi$$
$$+ \frac{3\kappa\mathfrak{e}}{16}(\overline{\psi}\gamma^5\gamma_a\psi)(\overline{\psi}\gamma^5\gamma^a\psi) \qquad (164)$$

being $\frac{3\kappa\mathfrak{e}}{16}(\overline{\psi}\gamma^5\gamma_a\psi)(\overline{\psi}\gamma^5\gamma^a\psi)$ as the repulsive 4-fermion interaction. By complex conjugation, we get

$$\psi^c = -i\gamma^2\psi^*, \qquad \psi^* = -i\gamma^2\psi^c, \qquad \gamma^{5*} = \gamma^5 \qquad (165)$$

Then, the Hehl-Datta equation takes the form

$$ie^\mu_a \gamma^a \psi^c_{;\mu} = m\psi^c + \frac{3\kappa}{8}(\overline{\psi^c}\gamma^5\gamma_a\psi^c)\gamma^5\gamma^a\psi^c \qquad (166)$$

[24] We are using the Einstein-Cartan-Sciama-Kibble (*ECSK*) theory of gravity, which naturally extends general relativity to include the intrinsic spin of matter [2-14]. In this theory, the spin of the Dirac fields is the source of the torsion tensor Sk, ij , which is the antisymmetric part of the affine connection $S^k_{i,j} = S_{kij} = Sk\,[ij]$. In turn, torsion, modifies the Dirac equation for elementary fermions by adding a cubic term in spinor fields, as shown by Hehl and Datta [15]. A nonlinear equation for fermions of this form was previously proposed by Heisenberg and Ivanenko [16, 17]. Such a term corresponds to an axial-axial 4-fermion interaction in the Lagrangian [7-15], generating a spinor-dependent, vacuum-energy term in the energy-momentum tensor [18-20].

Comparing to (163), we see opposite signs in the terms $-\frac{3\kappa}{8}(\overline{\psi}\gamma^5\gamma_a\psi)\gamma^5\gamma^a\psi$ and $+\frac{3\kappa}{8}(\overline{\psi^c}\gamma^5\gamma_a\psi^c)\gamma^5\gamma^a\psi^c$, suggesting the characterization of particles and anti-particles. Using the plane wave approximation, we obtain different dispersion relations between particles/anti-particles at finite densities and assuming a constant background torsion (see table 1).

However, the Lagrangian is C[25], that is to say, symmetric to charge conjugation, while the Hehl-Datta equation is not

$$\overline{\psi}^c\psi^c = -\overline{\psi}\psi,$$
$$(\overline{\psi}^c\gamma^\mu\gamma^5\psi^c)(\overline{\psi}^c\gamma_\mu\gamma^5\psi^c) = (\overline{\psi}\gamma^\mu\gamma^5\psi)(\overline{\psi}\gamma_\mu\gamma^5\psi) \qquad (167)$$

Fortunately, only chiral matter (e.g. right-handed neutrinos) can form condensates $\langle\overline{\psi}\gamma^5\gamma^0\rangle \equiv N \neq 0$. The inverse normalization of the Dirac spinor wave function of the order of (energy scale)^3, implies $N \approx T^3$ for the universe at temperature T. Thus, we have an abundance of neutrinos in the early universe, different to those of antineutrinos if $B^0 \neq 0$ being constant in a local frame, where we have the differences

[25] CPT symmetry is a fundamental principle of invariance or symmetry of physical laws. It establishes, under simultaneous transformations involving the inversion of the electrical charge, parity, and the sense of the time, with the temporal evolution equations of a physical process and the equations of an analogous process, with:
i) Charge conjugation C: All particles are substituted by their corresponding anti-particles.
ii) Parity inversion P: The parity of the process is inversed (with the interchange from right to left and with the change in the spin of the particles).
iii) Temporal inversion T: The sense of time is inversed.
These are invariant, described by the same equations, and give us the same results. The result of the last one is known as the CPT theorem, which affirms that all the quantum theory of local fields showing Lorentz covariance and defined by the Hermitian Hamiltonian, which is invariant under a transformation of CPT.

$$n_X - n_{\tilde{X}} = \frac{g}{(2\pi)^3}\int d^3p\left(\frac{1}{1+e^{E_X/T}} - \frac{1}{1+e^{E_{\tilde{X}}/T}}\right) \qquad (168)$$

which suggests that particle/anti-particle asymmetry can thus be generated from torsion. Could this be correct?

4.5.2. Spares

In the study of scalar fields in the early universe, extended particle physics models provide us with plenty of components and various effects that can help us use a string theory model as our model, including axions, inflatons, dilatons, and moduli stacks.

Some examples are linear in cosmic time, as with the background *KR*-pseudo-scalar $b \approx t$. Here, *b*-induced torsion produces matter/anti-matter asymmetry. The Logarithmic dilaton $\phi \approx \mathrm{Ln}a(t) = \mathrm{Ln}(t)$ also appears, which implies a linearly-expanding universe.

The various effects could be the result of the dilution of dark matter relic density due to coupling with running dilatons (Lahanas, NEM, Nanopoulos, Bender, and Sarkar).

TABLE 3: Different Dispersion Relations between Particles/Anti-Particles

Particle	Fermion	Anti-Fermion
f, f^{-1}	$\omega = m + \alpha\kappa N$	$\omega = m - \alpha\kappa N$

Another effect is obtained by the condensation of gravitino fields, which implies dynamical SUGRA breaking and suggests the existence of the inflaton. This is a small-field inflaton caused by the flat one-loop effective potential of the gravitino condensate (σ = *inflaton*, see fig. 26) (Ellis, NEM, Alexandre, and Houston).

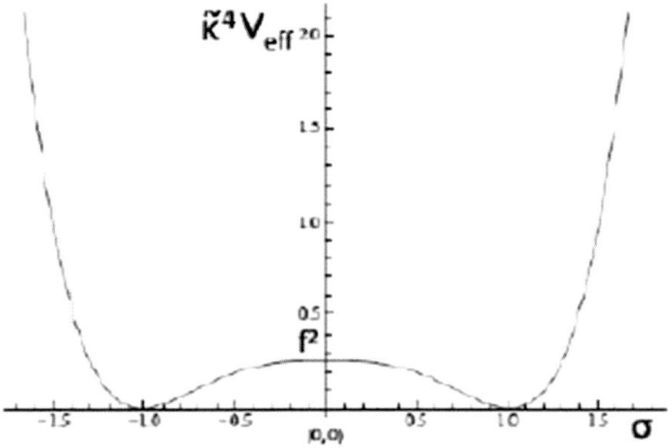

Fig. 26. Flat one-loop effective potential of gravitino condensate *versus* inflaton.

Some modifications are can be made according to the neutrinoless 2-beta decay rate with two flavor mixing (due to CPTV-modified effective mass) (M Sinha & B. Mukhopadhyay arXiv: 0704.2593). This happens if a small B^0 is present. If we ignore the neutrino/anti-neutrino mixing in this modification of the neutrinoless 2-beta decay rate (and considering small B^0), then we consider the amplitude $A \propto \sqrt{B_0^2 + m_e^2}$ and the small B^0 that exists today will also affect flavor oscillation. For example, we can consider the two flavors given by the Majorana field

$$\Psi = \begin{pmatrix} \psi_L^c \\ \psi_L \end{pmatrix}$$

then

$$\nu_{e,\mu} = \mathcal{U}^{\dagger}_{e,\mu} \Psi_{e,\mu} \qquad (169)$$

where

$$\mathcal{U}_{e,\mu} = \begin{pmatrix} \cos\theta_{e,\mu} & -\sin\theta_{e,\mu} \\ \sin\theta_{e,\mu} & \cos\theta_{e,\mu} \end{pmatrix} \qquad (170)$$

representing neutrino/anti-neutrino mixing on top of flavor mixing.

We can thus ask what flavor oscillations arise in the presence of gravity?

There are two sets of flavor eigenstates

$$\begin{aligned} \left|\nu_{e_1}\right\rangle &= \cos\phi_1\left|\chi_1\right\rangle - \sin\phi_1\left|\chi_2\right\rangle \\ \left|\nu_{e_2}\right\rangle &= \sin\phi_1\left|\chi_1\right\rangle + \cos\phi_1\left|\chi_2\right\rangle \end{aligned} \quad (171)$$

and

$$\begin{aligned} \left|\nu_{e_2}\right\rangle &= \cos\phi_2\left|\chi_3\right\rangle - \sin\phi_2\left|\chi_4\right\rangle \\ \left|\nu_{e_2}\right\rangle &= \sin\phi_2\left|\chi_3\right\rangle + \cos\phi_2\left|\chi_4\right\rangle \end{aligned} \quad (172)$$

For each set of flavor eigenstates, we have

$$\mathcal{P}_{fg_1} = \sin^2 2\phi_1 \sin^2 \delta_{fg_1}(t), \qquad \delta_{fg_1}(t) = \frac{\left|M_1^2 - M_2^2\right|}{4E} t, (173)$$

and

$$\mathcal{P}_{fg_2} = \sin^2 2\phi_1 \sin^2 \delta_{fg_2}(t), \qquad \delta_{fg_2}(t) = \frac{\left|M_3^2 - M4_2^2\right|}{4E} t, (174)$$

with oscillation lengths

$\lambda_{fg_1} = \dfrac{4\pi E}{\left|M_1^2 - M_2^2\right|}$ and $\lambda_{fg_2} = \dfrac{4\pi E}{\left|M_3^2 - M_4^2\right|}$, respectively.

Then ΔM^2 gives

$$\Delta M^2 = \left|M_1^2 - M_2^2\right| = \left|M_3^2 - M_4^2\right| = \left(\sqrt{B_0^2 + m_\mu^2} + \sqrt{B_0^2 + m_e^2}\right) \times$$
$$\times \sqrt{\left\{\left(\sqrt{B_0^2 + m_\mu^2} + \sqrt{B_0^2 + m_e^2}\right)^2 + 4m_{e\mu}^2\right\}} \quad (175)$$

These conditions are considered very stringent today due to the non-observation of neutrino/anti-neutrino oscillations.

4.5.3. First Results: Neutrino Interactions

As mentioned previously, the gravitational background generated CPT-violating effects in the early universe. In fact, only for this step of the universe's development can we establish theoretical models on the scattering differences provoked by the neutrino/anti-neutrino interactions. Thus, we can have different populations implying a particle freeze out and from which the baryogenesis of leptogenesis suggests baryogenesis (*baryogenesis/leptogenesis GUT or sphaleron*).

As such, CPTV effects different scenarios or *space-time-curvature/spin* couplings between neutrinos and antineutrinos [20] and coupling *curvature* to *fermion spin* may lead to different dispersion relations between neutrinos and antineutrinos (assuming dominance in the early eras) in the non-spherically symmetric geometries of the early universe.

A study realized from a point of view of the mathematical physics and particle physics carries the conjecture that we should consider the fermion particle along with the gauge boson for torsion, which is called the broson.

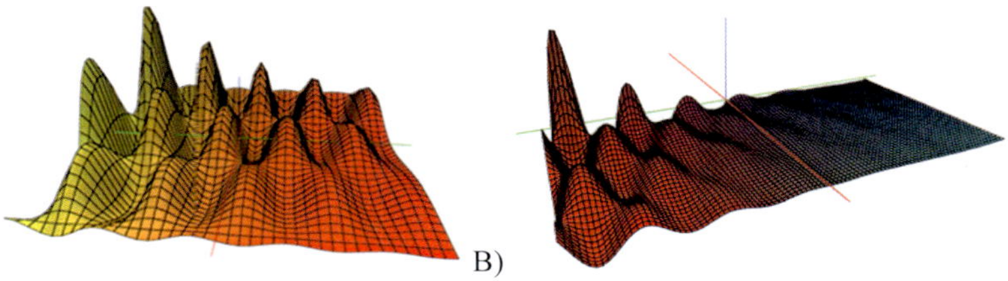

A) B)

Fig. 27. A) **H**-state superposition; B) 2-D surface model of curvature energy/fermion spin due to deformation, which could be due to potential generation for b-mixing with axions of dark matter. The curvature is an **H**-torsion.

Conjecture 4.5.3.1 (F. Bulnes, M. Ramírez, L. Ramirez, and O. Ramírez). The macroscopic image of the broson's actions must be a flow electro-gravitational energy whose micro-states begin from the actions of non-Abelian fields that are measured by the gauge fields.

The value of the broson is the intrinsic geometry of the torsion as a field observable, associated in a bundle. In field theory, it can be considered a framework operator belonging to certain algebras (such as quantum algebras) that comply with certain conditions for the solution of the field

$$(\Box + \chi T)o_{\mu}^{a} = 0 \qquad (176)$$

using **H**-states as the basis (states or densities in a Hamiltonian manifold, not necessarily **H**-particles) [14]. Likewise, the Dirac equation can take the following form

$$d\mathbf{h} = 0 \qquad (177)$$

$\forall \mathbf{h} \in \mathbf{H}$. The gauge bosons produce torsion in microscopic space due the electromagnetic characteristic of these bosons as photons [15], realizing backreactions with the space covered by the gravity. As a result, pending proof, we present the following theorem.

Theorem (F. Bulnes) 1. *Considering space-time with CPTV effects, the H-torsion is the deformation energy between neutrinos and antineutrinos or curvature/fermion spin energy (space-time-curvature/spin couplings between fermions/anti-fermions).*

Proof of Theorem 1. The CPTV effects of different space-time-curvature/spin couplings between fermions and anti-fermions give corresponding spinor representations of complex linear groups $GL_n(\mathbb{C})$, proving that solutions of the field equations are components for each *j*-th connection of the H-states [28][26]. They are determined by the corresponding hypercohomological orthogonal projection of these connections, being a consequence of the hypercohomological image in the category of complex vector spaces [29]. Alternatively, using deformation theory, the category of $\mathcal{M}$ of the $\mathcal{D}-$modules on the corresponding moduli stack Bun_G (the space of all bundles corresponding to the connections of the field equations)

[26] From a more general point of view (in mathematics), representations of these H-states can be made through spinor representations.

admits global presentation (for all particles/fields)

$$(\mathcal{D} \otimes \mathcal{K}^{1/2})^{\oplus_p} \mapsto (\mathcal{D} \otimes \mathcal{K}^{-1/2})^{\oplus_q} \mapsto \mathcal{M} \quad (178)$$

where $\mathcal{K}$ is the canonical line bundle on the moduli stack. The appearance of a square root $\mathcal{K}^{1/2}$ (related to the fact that the modules correspond to the deformed Lie algebra of the spectrum of Lie algebra of $GL_n(\mathbb{C})$) makes it more convenient to switch to the category of projective $\mathcal{D}-$modules with the same curvature as $\mathcal{K}^{1/2}$ and the category of spin $\mathcal{D}-$modules results. This is the category of fermions/anti-fermions whose "interacting curvature" can be space-time-curvature/spin couplings between fermions and anti-fermions. The corresponding category of space-time for this case can be defined by the sheaves of modules $\mathcal{O}$ rather than the twisted version

$$\mathcal{D}^s = \mathcal{K}^{-1/2} \otimes_{\mathcal{O}} \mathcal{D} \otimes_{\mathcal{O}} \mathcal{K}^{1/2} \quad (179)$$

We can undertake a major digression on the $\mathcal{D}-$modules that appear here, but this is not important to the current study. Here, the importance lies in the use of the "spin bundles" $\mathcal{K}^{1/2}$ or $\mathcal{K}^{-1/2}$, the corresponding Pfaffian case of the field equations (as described in (56)), which we wish to relate the **H**-states through $d\star\mathbf{H} = 0$ for the **H**-torsion (as a universal adjoint bundle over Bun_G). This arises from the choice of a spin structure with a smooth projective complex curve Σ (the particle trace in physical terms).

The general integral for the differential equation of the second order $d\mathbf{h} = 0,\ \forall\,\mathbf{h} \in \mathrm{H}^0(\Sigma, \mathrm{K}_\Sigma^{\otimes 2})$ (Higgs fields), is a superposing of the **H**-states.

In the case of spinor representation, the corresponding **H**-states can be given as spinor waves (see fig. 27 B)), which can be consigned to oscillations in the space-time-curvature/spin and microscopic deformation measured in $\mathcal{H}$ which is a deformed category

$$\mathcal{H} = H^{0}(\omega_{C}) \oplus H^{0}(\omega_{C}^{\otimes 2}) \oplus \ldots \oplus H^{0}(\omega_{C}^{\otimes n}) \qquad (180)$$

constructed by the superposition of **H**-states defined through their Hamiltonian densities [30]. (For more details of the categories, theories, results, and tools used in this demonstration, see [16, 29].)

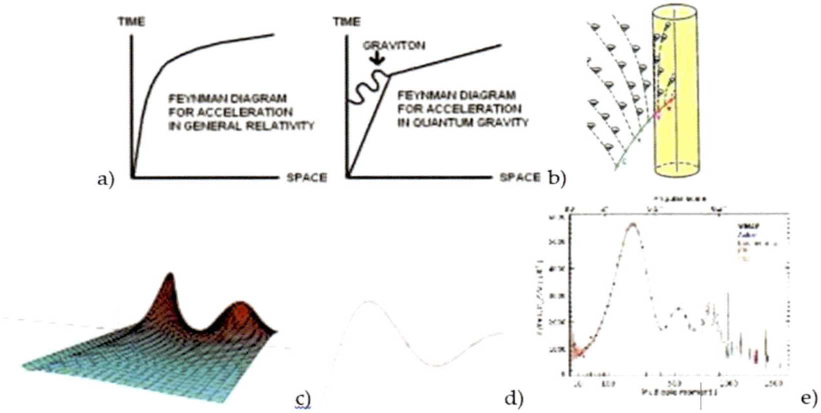

Fig. 28. a) The curvature can be measured to quantum level [8] as distortions given by the link-wave between a hypothetical particle like a dilaton (gauged graviton) and the trace on the relativistic Feynman diagram following quantum gravity [5, 8]. b) In the case of a cylindrical trajectory, for different times and considering the causality structure of space-time, the torsion can be determined by different deviations to the worldlines in each case having microscopic oscillations predicted by the Majorana fermions [16]. c) [15] Microscopic perturbation on a cylindrical surface. Also considered is the causal structure given by light cones. The red segment in figure 28 b) corresponds to the surface model given in 28 c). d) Curve of the cylindrical waves. e) Observation of similar waves of the spectrum power of CMB radiation temperature anisotropy in terms for 90o up to 0.4o to the multi-pole moment. Remember the relation of **H**-singularities with the field torsion in terms of large production or decay of energy.

4.5.4. Majorana Fermion Energies

The Dirac Lagrangian corresponding to the Majorana fields in this interaction is given by

$$\mathcal{L}=\sqrt{-g}(i\bar{\psi}\gamma^a D_a\psi - m\bar{\psi}\psi) \qquad (181)$$

where derivative D_a (*gravitational covariant derivative*) is given as (67) with the same coordinate system a. This includes the spin connection $\sigma^{ab}=\frac{1}{2}[\gamma^a,\gamma^b]$ and the proper gravitational connection is

$$\omega_{bca}=e_{b\lambda}(\partial_a e_c^\lambda + \Gamma^\lambda_{\gamma\mu} e_c^\gamma e_a^\mu) \qquad (182)$$

Then, the Lagrangian given in (181) can be rewritten as the sum of two Lagrangian components, one for the fermion and the other including the Majorana states

$$\mathcal{L}=\mathcal{L}_f+\mathcal{L}_I=\sqrt{-g}\bar{\psi}[(i\gamma^a\partial_a - m)+\gamma^a\gamma^5 B_a]\psi \qquad (183)$$

where B^d takes the form given in (70) and, for the Majorana neutrinos, above $\mathcal{L}_I$ is given by (71).

The term $\gamma^a\gamma^5 B_a$ from (183) comes from the Standard Model extension with Lorentz-violating coupling [21]. In relation to homogeneous and isotropic Friedman-Robertson-Walker geometries, the resulting Bianchi component B_a is annulled. This component can be constant in a given local frame in the early universe having axially symmetric Bianchi cosmologies or near rotating black holes.

Thus, we have the Bianchi cosmologies [22] below

$$ds^2=-dt^2+S(t)^2dx^2+R(t)^2[dy^2+f(y)^2dz^2]- \\ S(t)^2h(y)[2dx-h(y)dz]dz \qquad (184)$$

where

$$f(y) = \{y, \sinh y, \sin y\} \tag{185}$$

$$h(y) = \{-y^2/2, -\cosh y, \cos y\} \tag{186}$$

with Bianchi components I

$$B^0 = \frac{S[-f^2R^2(hf'R + Sh') + h^2S^2(hf'R + Sh')}{f^4R^4 + fh^2R^2S^2} \times$$

$$\times \frac{2fhRS(Rf' - hh'S)]}{f^4R^4 + fh^2R^2S^2} \tag{187}$$

and

$$B^2 = \frac{h[-f^2R^2 + 2fRS + h^2S^2][RS' - R'S]}{f^4R^4 + fh^2R^2S^2} \quad (188$$

Thus, for Bianchi II

$$B^0 = \frac{4R^3S + 3y^2RS^3 - 2yS^4}{8R^4 + 2y^2R^2S^2} \tag{189}$$

$$B^0 = \frac{(4yR^2 - 8RS - y^3S^2)(RS' - R'S)}{8R^4 + 2y^2R^2S^2} \tag{190}$$

The shift from torsion to spiralization of some particles is implicit in the equation.

For Bianchi VIII

$$B^0 = \frac{S[2\cosh^2 y(\cos 2y - 3)RS^2 - 4\cosh^2 y}{4(\cosh^2 y \sinh^2 y R^2S^2 + R^4 \sin^4 y)} \times$$

$$\times \frac{\sinh y S^3 + 4R^3 \cosh^2 y \sinh^2 y - R^2S(5\sinh y + \sinh 3y)}{4(\cosh^2 y \sinh^2 y R^2S^2 + R^4 \sin^4 y)} \tag{191}$$

and

$$B^2 = \frac{\cosh y(S^2\cosh^2 y + 2RS\sinh y - R^2\sinh^2 y)(RS' - R'S)}{\cosh^2 y\sinh yR^2S^2 + R^4\sinh^3 y}$$

(192)

For Bianchi IX

$$B^0 = \frac{S[2\cos^2 y(3-\cos 2y)RS^2 - 4\cos^2 y}{4(\cosh^2 y\sinh^2 yR^2S^2 + R^4\sin^4 y)}\times$$
$$\times\frac{\sin yS^3 - 4R^3\cos^2 y\sin^2 y + R^2S(5\sin y + \sin 3y)}{4(\cos^2 y\sin^2 yR^2S^2 + R^4\sin^4 y)}$$

(193)

and

$$B^2 = \frac{\cos y(S^2\cos^2 y + 2RS\sin y - R^2\sin^2 y)(RS' - R'S)}{\cos^2 y\sin yR^2S^2 + R^4\sin^3 y}$$

(194)

Then, we consider, as an example of the gravitational background generating CPT-violating effects, the neutrino/anti-neutrino asymmetry around black holes [23]. Considering the Kerr metric of a (rotating) black hole

$$ds^2 = \eta_{ij}dx^i dx^j - \left[\frac{2\alpha}{\varrho}s_i\upsilon_j + \alpha^2\upsilon_i\upsilon_j\right]dx^i dx^j, \quad (195)$$

where the mass relation M and the angular momentum relation a^2 are

$$\alpha = \frac{\sqrt{2Mr}}{\varrho}, \quad \varrho^2 = r^2 + \frac{a^2z^2}{r^2} \quad (196)$$

with $\upsilon_i = \left(1, \frac{ay}{a^2+r^2}, -\frac{ax}{a^2+r^2}, 0\right)$.

Being

$$s_i = \left(1, \frac{rx}{\sqrt{a^2 + r^2}}, \frac{ry}{\sqrt{a^2 + r^2}}, \frac{z\sqrt{r^2 + a^2}}{r}\right) \tag{197}$$

we have

$$r^4 - r^2(x^2 + y^2 + z^2 - a^2) - a^2 z^2 = 0 \tag{198}$$

Calculating vector B^{μ}, we can show that it does not vanish and hence CPT-violation is induced in this case.

Finally, considering the component B^0 as rotational, we can simplify the writing with new variables

$$B^0 = e_{1\lambda}(\partial_3 e_2^{\lambda} - \partial_2 e_3^{\lambda}) + e_{2\lambda}(\partial_1 e_3^{\lambda} - \partial_2 e_1^{\lambda}) + \\ + e_{3\lambda}(\partial_2 e_1^{\lambda} - \partial_1 e_2^{\lambda}) = -\frac{4a\sqrt{Mz}}{\bar{\varrho}^2\sqrt{2r^3}} \tag{199}$$

where $\bar{\varrho}^2 = 2r^2 + a^2 - x^2 - y^2 - z^2$.

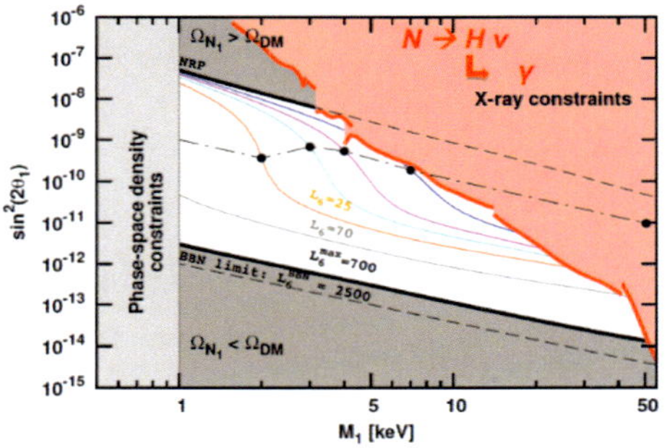

Fig. 29. Difference of light neutrinos, forecast from the Bianchi identity. The model is consistent with BBN and structure formation data in the universe and all other astrophysical constraints [31].

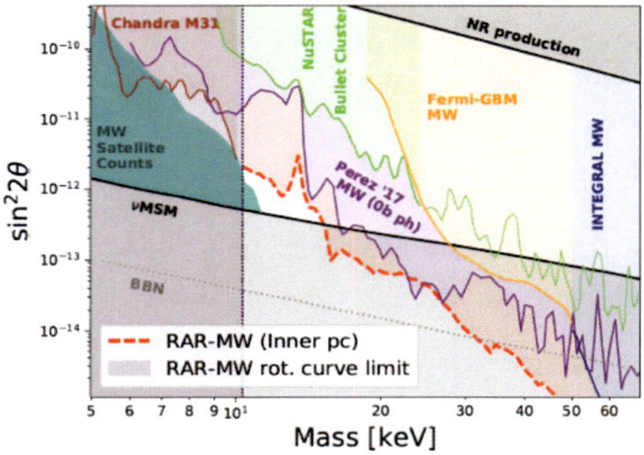

Fig. 30. The production of NR-MSM differences among neutrinos. Compatible with RHN as the addition of warm DM. This also satisfies the BH hypothesis and the galactic structure [31].

If we consider the neutrino/anti-neutrino population difference, we have

$$\Delta n = \frac{g}{(2\pi)^3}\int_{R_i}^{R_f} dV \int d^3|p|\left[\frac{1}{1+\exp(E_\nu/T)} - \frac{1}{1+\exp(E_{\bar{\nu}}/T)}\right] \quad (200)$$

Then, in relation to Kerr black holes, for which we have $B \bullet p << B^0 p_0$, we can demonstrate that

$$\Delta n = \frac{g}{(2\pi)^3}\int_{R_i}^{R_f}\int_0^\infty\int_0^\pi \left[\frac{1}{1+e^u\exp(B^0/T)} - \frac{1}{1+e^u\exp(-B^0/T)}\right] u^2 d\theta du dV$$

(201)

with $u = |p|/T$ then, if $B << T$ we have

$$\Delta n \approx gT^3\left(\frac{\overline{B}^0}{T}\right) \quad (202)$$

where warm DM adding B^0 is averaged over the spatial volume V. The dispersion relations of neutrinos are different from those of anti-neutrinos in such geometries, if B^{μ}, is constant in the local frame. If, for the Dirac equations, we consider the mass as

$$(\boldsymbol{p}_a \pm \boldsymbol{B}_a)^2 = m^2 \qquad (203)$$

where $\pm$ refers to the critical fields (here neutrino/anti-neutrino), then we have the following energy relations (*CPTV scattering relations*)

$$\boldsymbol{E} = \sqrt{(\boldsymbol{p}_a - \boldsymbol{B}_a)^2 + m^2} + \boldsymbol{B}_0 \qquad (204)$$

and their conjugate is

$$\overline{\boldsymbol{E}} = \sqrt{(\boldsymbol{p}_a - \boldsymbol{B}_a)^2 + m^2} - \boldsymbol{B}_0 \qquad (205)$$

However, the masses are equal between the particle and anti-particle sectors. The abundance of neutrinos in the early universe is different the case of of anti-neutrinos if $B^0 \neq 0$ and is constant in the local frame. Therefore, the difference can established by

$$\Delta n \approx g^* T^3 \left(\frac{B^0}{T} \right) \qquad (206)$$

where g^* is the degrees of freedom for the (relativistic) neutrino.

Looking at $\Delta n \approx g T^3 \left(\frac{\overline{B}^0}{T} \right)$, we can see that the asymmetry depends on the sign of B^0.

Primordial black holes, with masses of $M_{BH} < 10^{15}$ gm no longer exist today and only black holes with masses of $M_{BH} > 10^{15}$ gm may currentlyexist . If we consider the Hawking temperature [24]

$$T = \frac{\hbar}{8\pi k_B M} \approx 10^{-7}\mathrm{K}\left(\frac{M_{\odot}}{M}\right) \tag{207}$$

where $T \approx 10^{11}\,\mathrm{K} \approx 1.6 \times 10^{-5}\,erg$ and

$$\overline{B}_0 \approx 1.6 \times 10^{-6}\,erg \tag{208}$$

then $\Delta n \approx 10^{-16}$. To reproduce the observed Baryon asymmetry $\Delta n = \mathrm{O}(10^{-10})$, we need $10^6\,\mathrm{BH}$, with the same sign of B^0 and having fine tuning.

4.5.5. Spinor Indices for Weyl Field Solutions: Photonic Evolution in Universal Processes Related to Torsion

4.5.5.1. Introduction

In the different wave solutions to the Weyl field equations, there exist many arrangements and permutation of indices that establish rules for the transformation of photons and lay out correspondences between photons and the Majorana and Dirac fields. This permits the conformation of an intertwining operator of the energy states of these fields. In this intertwining operator, we see photon transformation established from fermion interactions. This gives us a fundamental understanding of the evolution of the photonic framework in processes related to torsion in the universe.

The spinor indices help us distinguish the generator matrices for the infinitesimal transformation L [61] and given by the spin matrices $(S_L^{\mu\nu})_a^b$ from the symbols for invariance and covariance given by $a, b, \mu,$ and $\nu,$ which establish the rules of transformation inside the kernels of the Weyl equation for different cameras of the root spaces [62]. These different images shape the spaces of different faces of the fermions that appear in the photon transformation.

With an order of invariance under the Lorentz transformation of the Dirac and Majorana fields

$$U(\Lambda)^{-1}\Psi(x)U(\Lambda) = D(\Lambda)\Psi(\Lambda^{-1}x) \qquad (209)$$

or their equivalents in the spinor frame

$$U(\Lambda)^{-1}\psi_a(x)U(\Lambda) = L(\Lambda)_a^c\psi_c(\Lambda^{-1}x) \qquad (210)$$

and

$$U(\Lambda)^{-1}\psi_{\dot{a}}^{\dagger}(x)U(\Lambda) = L(\Lambda)_{\dot{a}}^{\dot{c}}\psi_{\dot{c}}^{\dagger}(\Lambda^{-1}x) \qquad (211)$$

Using the corresponding Weyl camera for a root space, we can establish the rules of Feynman diagrams that give new infinitesimal transformations useful for unifying the photon sources required in fermion interactions to annihilate or create photons with the properties of a universal source, as boson gauges or Higgs boson coverings.

This section presents the different transformations involved inside infinitesimal transformations like the one given by $U(\Lambda)^{-1}\Psi(x)U(\Lambda)$, due to the indices of tensor expression and their spinors.

We also analyze some correlations between the Majorana and Dirac fields to characterize all the infinitesimal transformations of all the fermion interactions, since these fermion phenomena are where all particles with charges are produced, including ghosts, which are traces of the existence of particles as neutrinos, filling all the space-time.

Finally, we wish to establish that the process of producing all photon types is equivalent to defining the problem of spinor technology in the photonic context and thus it will be necessary to extend the concept of photon sources to only one universal source of photons.

4.5.5.2. Considering Some Meanings of Spinor Indices

We consider the spinor given by the wave-field solution

$$\psi^a(x) \equiv \varepsilon^{ab}\psi_b(x) \qquad (212)$$

where the invariant symbols for raising and lowering the spinor indices a and b satisfy

$$\varepsilon^{ab} = -\varepsilon_{ab} = i\sigma_2 \qquad (213)$$

and in which σ_2 can be the matrices

$$\sigma_1 = \begin{pmatrix} 0 & 1 \\ 1 & 0 \end{pmatrix}, \quad \sigma_2 = \begin{pmatrix} 0 & -i \\ i & 0 \end{pmatrix}, \quad \sigma_3 = \begin{pmatrix} 1 & 0 \\ 0 & -1 \end{pmatrix}$$

In particular, we consider

$$\varepsilon^{12} = \varepsilon^{\dot{1}\dot{2}} = \varepsilon_{21} = \varepsilon_{\dot{2}\dot{1}} = +1 \qquad (214)$$

and also the case

$$\varepsilon^{21} = \varepsilon^{\dot{2}\dot{1}} = \varepsilon_{12} = \varepsilon_{\dot{1}\dot{2}} = -1 \qquad (215)$$

We can generate other invariant symbols considering the identy matrix I and σ^{μ}, having

$$\sigma^{\mu}_{a\dot{a}} = (I, \sigma^{\mu}) \qquad (216)$$

we generate field objects, such that

$$A_{a\dot{a}}(x) = \sigma^{\mu}_{a\dot{a}} A_{\mu}(x) \qquad (117)$$

Simple identities with $\sigma^{\mu}_{a\dot{a}}$ are obtained

$$\sigma^{\mu}_{a\dot{a}} \sigma_{\mu b\dot{b}} = -2\varepsilon_{ab}\varepsilon_{\dot{a}\dot{b}} \qquad (218)$$

and

$$\varepsilon^{ab}\varepsilon^{\dot{a}\dot{b}}\sigma^{\mu}_{a\dot{a}}\sigma^{\nu}_{b\dot{b}} = -2\varepsilon_{ab}\varepsilon_{\dot{a}\dot{b}} \tag{219}$$

which are proportionally constant from direct calculation.

For an infinitesimal transformation L, we can consider the identity

$$\varepsilon_{ab} = L(\Lambda)^{c}_{a} L(\Lambda)^{d}_{b}\varepsilon_{cd} \tag{220}$$

having $\Lambda^{\mu}_{\nu} = \delta^{\mu}_{\nu} + \delta\omega^{\mu}_{\nu}$ and then (220) takes the form

$$L^{b}_{a}(1+\delta\omega) = \delta^{b}_{a} + \frac{i}{2}\delta\omega_{\mu\nu}(S^{\mu\nu}_{L}), \tag{221}$$

and for tensor ε_{ab}, given in (220), we find

$$\varepsilon_{ab} = \varepsilon_{ab} + \frac{i}{2}\delta\omega_{\mu\nu}[(S^{\mu\nu}_{L})^{c}_{a}\varepsilon_{cb} + (S^{\mu\nu}_{L})^{d}_{b}\varepsilon_{ad}] + O(\delta\omega^{2})$$
$$= \varepsilon_{ab} + \frac{i}{2}\delta\omega_{\mu\nu}[-(S^{\mu\nu}_{L})_{ab} + (S^{\mu\nu}_{L})_{ba}] + O(\delta\omega^{2})$$

From $[-(S^{\mu\nu}_{L})_{ab} + (S^{\mu\nu}_{L})_{ba}]$, which implies that $(S^{\mu\nu}_{L})_{ab} = (S^{\mu\nu}_{L})_{ba}$ with similarity

$$(S^{\mu\nu}_{R})_{\dot{a}\dot{b}} = (S^{\mu\nu}_{R})_{\dot{b}\dot{a}} \tag{222}$$

where R is the infinitesimal transformation given by $\sigma^{\mu}_{a\dot{a}}$, we consider the identity of these infinitesimal transformations R

$$\sigma^{\rho}_{a\dot{a}} = \Lambda^{\rho}_{\tau} L(\Lambda)^{b}_{a} R(\Lambda)^{\dot{b}}_{\dot{a}}\sigma^{\tau}_{b\dot{b}} \tag{223}$$

Newly using the expressions

$$\Lambda^{\rho}_{\tau} = \delta^{\rho}_{\tau} + \frac{i}{2}\delta\omega_{\mu\nu}(S^{\mu\nu}_{V}) \qquad (224)$$

$$L^{b}_{a}(1+\delta\omega) = \delta^{b}_{a} + \frac{i}{2}\delta\omega_{\mu\nu}(S^{\mu\nu}_{L})^{b}_{a} \qquad (225)$$

$$R^{\dot{b}}_{\dot{a}}(1+\delta\omega) = \delta^{\dot{b}}_{\dot{a}} + \frac{i}{2}\delta\omega_{\mu\nu}(S^{\mu\nu}_{L})^{\dot{b}}_{\dot{a}} \qquad (226)$$

where, in (224), we have used the spin matrix expression given by

$$(S^{\mu\nu}_{V})^{\rho}_{\tau} \equiv \frac{1}{i}(g^{\mu\rho}\delta^{\nu}_{\tau} - g^{\nu\rho}\delta^{\mu}_{\tau}) \qquad (227)$$

Isolating linear terms in $\delta\omega_{\mu\nu}$, we have

$$(g^{\mu\rho}\delta^{\nu}_{\tau} - g^{\nu\rho}\delta^{\mu}_{\tau})\sigma^{\tau}_{a\dot{a}} + i(S^{\mu\nu}_{L})^{b}_{a}\sigma^{\rho}_{b\dot{a}} + i(S^{\mu\nu}_{R})^{\dot{b}}_{\dot{a}}\sigma^{\rho}_{a\dot{b}} = 0, (228)$$

Multiplying by $\sigma_{\rho c\dot{c}}$, we have

$$\sigma^{\mu}_{c\dot{c}}\sigma^{\nu}_{a\dot{a}} - \sigma^{\nu}_{c\dot{c}}\sigma^{\mu}_{a\dot{a}} + i(S^{\mu\nu}_{L})^{b}_{a}\sigma^{\rho}_{b\dot{a}} + i(S^{\mu\nu}_{R})^{\dot{b}}_{\dot{a}}\sigma^{\rho}_{a\dot{b}}\sigma_{\rho c\dot{c}} = 0, (229)$$

Using the fact that $\sigma^{\mu}_{a\dot{a}}\sigma_{\mu b\dot{b}} = -2\varepsilon_{ab}\varepsilon_{\dot{a}\dot{b}}$, equation (229) takes the form

$$\sigma^{\mu}_{c\dot{c}}\sigma^{\nu}_{a\dot{a}} - \sigma^{\nu}_{c\dot{c}}\sigma^{\mu}_{a\dot{a}} + 2i(S^{\mu\nu}_{L})_{ac}\varepsilon_{\dot{a}\dot{c}} + 2i(S^{\mu\nu}_{R})_{\dot{a}\dot{c}}\varepsilon_{ac} = 0, (230)$$

Multiplying the matrices (*left and right*) $(S^{\mu\nu}_{L})_{ac}$ and $(S^{\mu\nu}_{R})_{\dot{a}\dot{c}}$ by $\varepsilon^{\dot{a}\dot{c}}$ and ε^{ac}, respectively, we obtain the expressions

$$(S_L^{\mu\nu})_{ac} = \frac{i}{4}\varepsilon^{\dot{a}\dot{c}}(\sigma^{\mu}_{a\dot{a}}\sigma^{\nu}_{c\dot{c}} - \sigma^{\nu}_{a\dot{a}}\sigma^{\mu}_{c\dot{c}}) \qquad (231)$$

$$(S_R^{\mu\nu})_{\dot{a}\dot{c}} = \frac{i}{4}\varepsilon^{ac}(\sigma^{\mu}_{a\dot{a}}\sigma^{\nu}_{c\dot{c}} - \sigma^{\nu}_{a\dot{a}}\sigma^{\mu}_{c\dot{c}}) \qquad (232)$$

in which it is necessary to observe that $\varepsilon^{\dot{a}\dot{c}}\varepsilon_{\dot{a}\dot{c}} = -2$ to (231) and $\varepsilon^{\dot{a}\dot{c}}(S_R^{\mu\nu})_{\dot{a}\dot{c}}$ [27] to (232).

Defining

$$\bar{\sigma}^{\mu\dot{a}a} \equiv \varepsilon^{ab}\varepsilon^{\dot{a}\dot{b}}\sigma^{\mu}_{b\dot{b}} \qquad (233)$$

where the invariant symbol $\bar{\sigma}^{\mu\dot{a}a}$ (as half-rotation given by the right rotation from (216)) is given by

$$\bar{\sigma}^{\mu\dot{a}a} = (I, -\sigma^{\mu}) \qquad (234)$$

Then, we find that $(S_L^{\mu\nu})_a^{\ b}$ and $(S_R^{\mu\nu})^{\dot{a}}_{\ \dot{b}}$ take the form

---

[27] Particular cases of the identities and Fierz relations:

$\varepsilon_{\alpha\beta}\varepsilon^{\gamma\delta} = -\delta^{\gamma}_{\alpha}\delta^{\delta}_{\beta} + \delta^{\delta}_{\alpha}\delta^{\gamma}_{\beta}$, $\quad \varepsilon_{\dot{\alpha}\dot{\beta}}\varepsilon^{\dot{\gamma}\dot{\delta}} = -\delta^{\dot{\gamma}}_{\dot{\alpha}}\delta^{\dot{\delta}}_{\dot{\beta}} + \delta^{\dot{\delta}}_{\dot{\alpha}}\delta^{\dot{\gamma}}_{\dot{\beta}}$,

$\sigma^{\mu}_{\alpha\dot{\alpha}}\bar{\sigma}_{\mu}^{\dot{\beta}\beta} = 2\delta^{\beta}_{\alpha}\delta^{\dot{\beta}}_{\dot{\alpha}}$, $\quad \sigma^{\mu}_{\alpha\dot{\alpha}}\sigma_{\mu\beta\dot{\beta}} = 2\delta\varepsilon_{\alpha\beta}\varepsilon_{\dot{\alpha}\dot{\beta}}$,

$\bar{\sigma}^{\mu\dot{\alpha}\alpha}\bar{\sigma}_{\mu}^{\dot{\beta}\beta} = 2\varepsilon^{\alpha\beta}\varepsilon^{\dot{\alpha}\dot{\beta}}$, $\quad [\sigma^{\mu}\bar{\sigma}^{\nu} + \sigma^{\nu}\bar{\sigma}^{\mu}]_{\alpha}^{\ \beta} = 2g^{\mu\nu}\delta^{\beta}_{\alpha}$,

$[\bar{\sigma}^{\mu}\sigma^{\nu} + \bar{\sigma}^{\nu}\sigma^{\mu}]^{\dot{\alpha}}_{\ \dot{\beta}} = 2g^{\mu\nu}\delta^{\beta}_{\alpha}$,

$\sigma^{\mu}\bar{\sigma}^{\nu} + \sigma^{\rho} = g^{\mu\nu}\sigma^{\rho} - g^{\mu\rho}\sigma^{\nu} + g^{\nu\rho}\sigma^{\mu} + i\varepsilon^{\mu\nu\rho\kappa}\sigma_{\kappa}$,

$\bar{\sigma}^{\mu}\sigma^{\nu}\bar{\sigma}^{\rho} = g^{\mu\nu}\bar{\sigma}^{\rho} - g^{\mu\rho}\bar{\sigma}^{\nu} + g^{\nu\rho}\bar{\sigma}^{\mu} + i\varepsilon^{\mu\nu\rho\kappa}\bar{\sigma}_{\kappa}$.

$$(S_L^{\mu\nu})_a{}^b = +\frac{i}{4}(\sigma^\mu \bar{\sigma}^\nu - \sigma^\nu \bar{\sigma}^\mu)_a{}^b \qquad (235)$$

$$(S_R^{\mu\nu})^{\dot{a}}{}_{\dot{b}} = -\frac{i}{4}(\bar{\sigma}^\mu \sigma^\nu - \bar{\sigma}^\nu \sigma^\mu)^{\dot{a}}{}_{\dot{b}} \qquad (236)$$

for which we can contract the indices in (235) and (236) giving a place to the symbols ${}^c{}_c$ and ${}^{\dot{c}}{}_{\dot{c}}$, respectively, which define the spinors to left-handed Weyl fields (in the fermion context)

$$\chi\psi = \chi^a\psi_a \qquad (237)$$

and

$$\chi^\dagger\psi^\dagger = \chi^\dagger_{\dot{a}}\psi^{\dagger\dot{a}} \qquad (238)$$

where these wave-fields correspond to spin 1/2 particles, that are anti-commuting fermions (spin-statistics theorem [63]) (see fig. 31)

$$\chi_a(x)\psi_b(y) = -\psi_b(y)\chi_a(x) \qquad (239)$$

From (237), we find

$$\chi\psi = \chi^a\psi_a = -\psi_a\chi^a = \psi^a\chi_a = \psi\chi \qquad (240)$$

where the invariance and covariance properties of the spinors remain simplified as ${}_a{}^a = {}^a{}_a$

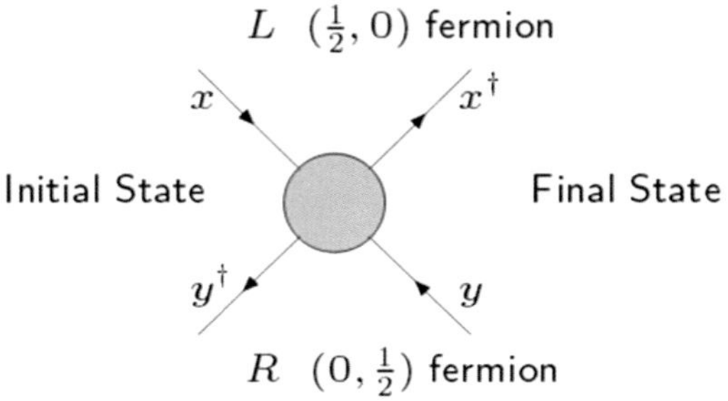

Fig. 31. The rules for assigning two-component external state spinors [64-66].

Then the 1/2-particles are fermions that anti-commute. Determining the Hermitian conjugate from (240), we find

$$(\chi\psi)^\dagger = (\chi^a \psi_a)^\dagger = (\psi_a)^\dagger (\chi^a)^\dagger = \psi^\dagger_{\dot{a}} \chi^{\dagger \dot{a}} = \psi^\dagger \chi^\dagger \qquad (241)$$

as is expected if we ignore the indices.

By similarity, we have

$$\chi^\dagger \psi^\dagger = \chi^\dagger \psi^\dagger \qquad (242)$$

and thus our right-handed field will always be written with the dagger operator.

If we consider the spin matrix transformation $\bar{\sigma}^\mu$, applied to the spinor χ, that is to say, of the type $\bar{\sigma}^\mu \chi$, we find that

$$\psi^\dagger \bar{\sigma}^\mu \chi = \psi^\dagger_{\dot{a}} \bar{\sigma}^{\mu \dot{a} c} \chi_c \qquad (243)$$

which is managed like a vector field under Lorentz transformations

$$U(\Lambda)^{-1}[\psi^{\dagger}\bar{\sigma}^{\mu}\chi]U(\Lambda)=\Lambda^{\mu}{}_{\nu}[\psi^{\dagger}\bar{\sigma}^{\nu}\chi] \qquad (244)$$

where $[\psi^{\dagger}\bar{\sigma}^{\nu}\chi]$ is evaluated in $\Lambda^{-1}x$. The Hermitian conjugate from (243) is

$$\begin{aligned}[\psi^{\dagger}\bar{\sigma}^{\nu}\chi]^{\dagger} &= [\psi^{\dagger}_{\dot{a}}\bar{\sigma}^{\mu a\dot{c}}\chi_{c}]^{\dagger}\\ &= \chi^{\dagger}_{\dot{c}}(\bar{\sigma}^{\mu a\dot{c}})^{*}\psi_{a}\\ &= \chi^{\dagger}_{\dot{c}}(\bar{\sigma}^{\mu\dot{c}a})^{*}\psi_{a}\\ &= \chi^{\dagger}\bar{\sigma}^{\mu}\psi\end{aligned}$$

where, in the equality step $\chi^{\dagger}_{\dot{c}}(\bar{\sigma}^{\mu\dot{c}a})^{*}\psi_{a}$, the transformation obtained by $\bar{\sigma}^{\mu a\dot{a}}=(I,-\sigma^{\mu})$, is Hermitian.

4.5.5.3. Majorana and Dirac Fields from Fermion Interactions

From the Lagrangian proposed in [67] and written as the Hermitian matrices $\bar{\sigma}^{\mu}$, we have

$$\mathcal{L}=i\chi^{\dagger}\bar{\sigma}^{\mu}\partial_{\mu}\chi+i\xi^{\dagger}\bar{\sigma}^{\mu}\partial_{\mu}\xi-m\chi\xi-m\xi^{\dagger}\chi^{\dagger},(245)$$

There is an additional discrete symmetry that exchanges the two fields χ_a and ξ_a, as[28]

[28] In fact it has the property $C^{-1}\mathcal{L}(x)C=\mathcal{L}(x)$.

$$C^{-1}\chi_a(x)C = \xi_a(x) \qquad (246)$$

and

$$C^{-1}\xi_a(x)C = \chi_a(x) \qquad (247)$$

where C, is the unitary charge conjugation operator. We wish to express (246) and (247) in terms of the Dirac and Majorana fields. As defined in [67], we have the charge conjugation matrix

$$C = \begin{pmatrix} \varepsilon_{ac} & 0 \\ 0 & \varepsilon^{\dot{a}\dot{c}} \end{pmatrix} \qquad (248)$$

Then, the Dirac field Ψ (the field from the Dirac equation $(-i\gamma^{\mu}\partial_{\mu} + m)\Psi = 0$) takes the form

$$\Psi = \begin{pmatrix} \chi_c \\ \xi^{\dagger\dot{c}} \end{pmatrix} \qquad (249)$$

and

$$\Psi^{\mathrm{C}} = C\overline{\Psi}^{\mathrm{T}} = \begin{pmatrix} \xi_a \\ \chi^{\dagger\dot{a}} \end{pmatrix} \qquad (250)$$

where ^{C} and ^{T} are conjugation and transposition transformations defined in [67]. Thus, we have

$$C^{-1}\Psi(x)C = \Psi(x) \qquad (251)$$

The charge conjugation matrix has following properties

$$C^{-1} = C^{\mathrm{T}} = C^{\dagger} = -C \qquad (252)$$

and (248) can also be written as

$$C = \begin{pmatrix} -\varepsilon^{ac} & 0 \\ 0 & -\varepsilon_{\dot{a}\dot{c}} \end{pmatrix} \qquad (253)$$

Then, we find useful the identity related to the Dirac matrix γ^{μ}, given in [61, 67]

$$C^{-1}\gamma^{\mu}C = -(\gamma^{\mu})^{\mathrm{T}} \qquad (254)$$

Thus, the Majorana field is properly conjugated from Ψ, giving us

$$\Psi^{\mathrm{C}} = \Psi \qquad (255)$$

where the explicated Majorana field is

$$\Psi = \begin{pmatrix} \psi_c \\ \psi^{\dagger \dot{c}} \end{pmatrix} \qquad (256)$$

and we have a similar property for the scalar fields $\varphi^{\dagger} = \varphi$.

Incorporating the Majorana condition, we have the Lagrangian (245)

$$\mathcal{L} = \frac{i}{2}\Psi^{\mathrm{T}}C\gamma^{\mu}\partial_{\mu}\Psi - \frac{1}{2}m\Psi^{\mathrm{T}}C\Psi \qquad (257)$$

which is the Majorana Lagrangian. If we wish to go back from the four-component Dirac or Majorana fields to the two-component Weyl fields, it is useful to define the projection matrix

$$\gamma_5 \equiv \begin{pmatrix} -\delta_a^{\ c} & 0 \\ 0 & +\delta^{\dot{a}}_{\ \dot{c}} \end{pmatrix} \qquad (258)$$

for which the Dirac matrix γ_5 is very relevant as it defines the universal transformation mechanisms that contribute to the creation and annihilation of the relevant particles. The anti-symmetries that appear in the interaction between anti-particles and particles involve this matrix in their transformation when considering Dirac fermions or Majorana fermions [61].

We can define the left and right projection matrices as

$$P_L = \frac{1}{2}(1-\gamma_5) = \begin{pmatrix} \delta_a{}^c & 0 \\ 0 & 0 \end{pmatrix} \tag{259}$$

and

$$P_L = \frac{1}{2}(1+\gamma_5) = \begin{pmatrix} 0 & 0 \\ 0 & \delta^{\dot{a}}{}_{\dot{c}} \end{pmatrix}, \tag{260}$$

and, for a Dirac field, we find

$$P_L\Psi = \begin{pmatrix} \chi_c \\ 0 \end{pmatrix}, \qquad P_R\Psi = \begin{pmatrix} 0 \\ \xi^{\dagger\dot{c}} \end{pmatrix}, \tag{261}$$

where Ψ is given as in (256).

4.5.5.4. Photon Transformation

We consider the requirement of total symmetry for a multi-spinor source $\psi_{\zeta_1\zeta_2\ldots\zeta_n}(x)$, which can be written as

$$\psi_{\zeta_1\zeta_2\ldots\zeta_n}(x) = \psi_{\zeta_{\alpha_1}\zeta_{\alpha_2}\ldots\zeta_{\alpha_n}}(x) \tag{262}$$

where $\alpha_1, \alpha_2, \ldots, \alpha_n$ is any permutation of $1, \ldots, n.$

The multi-spinor refers to a larger system than desired and has projection matrices as given in (259) and (260) in the corresponding range. Indeed, it is sufficient to use such spin 1/2-projection matrices on each spinor index to obtain the required reduction to a physical system of spin $s = \frac{1}{2}n,$ for a symmetrical multi-spinor. As defined in (259) and (260), the rest frame of a massive particle is the projection matrix given by

$$\prod_{\alpha=1}^{n} \frac{1}{2}(1+\gamma^0)_\alpha \tag{263}$$

where α designates the spinor index on which the corresponding matrix acts.

In our previous exposition (section 4.5.5.3), we considered $n = 2$ and chose as the anti-symmetrical spinor

$$\psi_{\zeta_1\zeta_2}(x) = \psi_{\zeta_2\zeta_1}(x) \tag{264}$$

According to our theory of indices, an anti-symmetrical function of a pair of effectively two-valued indices has only one independent component. The general expression of these observations is given by the following vacuum amplitude [61]

$$\left\langle 0_+ \middle| 0_- \right\rangle^S = \exp\{iW(\Psi)\} \tag{265}$$

where, using 4-dimensional momentum space, we have

$$W(\Psi) = \frac{1}{2}\int \frac{(dp)}{(2\pi)^4} \Psi(-p) \prod_{\alpha}^{n} [\gamma^0(m - \gamma p)]_\alpha \times$$
$$\times \left[\frac{1}{p^2 + m^2 - i\varepsilon}\right] \Psi(p) \tag{266}$$

The kernel of this quadratic form is given by $\prod_{\alpha}^{n} [\gamma^0(m - \gamma p)]_\alpha$ and has a defined symmetry under the matrix transposition combined with the substitution $p^\mu \to -p^\mu$, thus

$$\prod_{\alpha}^{n}[\gamma^{0}(m+\gamma p)]_{\alpha}^{\mathrm{T}}=(-1)^{n}\prod_{\alpha}^{n}[\gamma^{0}(m-\gamma p)]_{\alpha} \qquad (267)$$

Using the algebraic properties of the photon source [61] and matching the symmetry properties of the kernel of transformation (*action*) given by (266), we have the photon statistics according to Boson-Einstein statistics and Fermion-Dirac statistics. As we wish to obtain corresponding photons for the Majorana and Dirac fields, we need a corresponding photon source to produce all the different 1/2-spin particles and neutrinos derived from these 1/2-spin particles.

To do this, we consider the vacuum amplitude (265) with the quantum transformation (action) from the leptons

$$\langle 0_{+}|0_{-}\rangle \rightarrow \exp\{\frac{1}{2}\int(dx)(dx')\eta(x)\frac{1}{2}(1\pm l\alpha_{5})\times$$
$$\times(-\alpha_{\mu}\partial_{\mu})\int d\omega_{p}e^{ip(x-x')}\eta'(x)\}_{Euclidean} \qquad (268)$$

This restricts us to the solutions of the Dirac equation

$$(1-i\lambda q\gamma_{5})\eta(x)=0 \qquad (269)$$

limiting us to the particle family with leptonic charges given by the condition $\lambda^{2}=1$ [4] where two kinds of particles are distinguished by the corresponding projection factor $\frac{1}{2}(1\pm il\gamma_{5})$.

Finally, according to the Lorentz transformation behavior

$$U(\Lambda)^{-1}\Psi(x)U(\Lambda)=D(\Lambda)\Psi(\Lambda^{-1}x)$$

of γ_{5}, it follows that $\gamma^{\mu}\gamma^{\nu}\gamma^{\kappa}\gamma^{\lambda}=e^{\mu\nu\kappa\lambda}\gamma_{5}$, as

$$L^{-1}\gamma_{5}L=l^{0}{}_{\mu}l^{1}{}_{\nu}l^{2}{}_{\kappa}l^{3}{}_{\lambda}e^{\mu\nu\kappa\lambda}\gamma_{5} \qquad (270)$$

We can obtain the two-component field for the fermions (see the table below).

TABLE I
FERMIONS AND ANTI-FERMIONS NAMES AND TWO-COMPONENT FIELDS

Fermion name	Two-Component Fields	$L^{-1}\gamma_5 L$ – Transformation Through Majorana Fermions
ℓ^- (Lepton)	$\ell^-, \ell^\dagger$	$\ell^{-\mu}{}_{\kappa}\ell^{-\nu}{}_{\lambda}\sigma^{\kappa\lambda}, (\cong \ell^\dagger)$
ℓ^+ (Anti-lepton)	$\ell^+, \ell^\dagger$	$\ell^{+}{}_{\mu}{}^{\kappa}\ell^{+}{}_{\nu}{}^{\lambda}\sigma^{\kappa\lambda}, (\cong \ell^\dagger)$
ν (neutrino)	$\nu, \bar{\nu}^\dagger$	For Majorana fermions defined through the Dirac fields given by (39) satisfies $\Psi_M = \Psi_M^c = C\Psi_M{}^T$, the following conditions are satisfied: $\Psi_{M_1}\Psi_{M_2} = \Psi_{M_2}\Psi_{M_1}$, $\Psi_{M_1}\gamma_5\Psi_{M_2} = \Psi_{M_2}\gamma_5\Psi_{M_1}$, $\Psi_{M_1}\gamma^\mu\Psi_{M_2} = -\Psi_{M_2}\gamma^\mu\Psi_{M_1}$, $\Psi_{M_1}\gamma^\mu\gamma_5\Psi_{M_2} = \Psi_{M_2}\gamma^\mu\gamma_5\Psi_{M_1}$, $\Psi_{M_1}\Sigma^{\mu\nu}\Psi_{M_2} = -\Psi_{M_2}\Sigma^{\mu\nu}\Psi_{M_1}$, $\Psi_{M_1}\Sigma^{\mu\nu}\gamma_5\Psi_{M_2} = -\Psi_{M_2}\Sigma^{\mu\nu}\gamma_5\Psi_{M_1}$.
$\bar{\nu}$ (Anti-neutrino)	$\bar{\nu}, \nu^\dagger$	
q (quark)	$q, \bar{q}^\dagger$	
$\bar{q}$ (Anti-quark)	$\bar{q}, q^\dagger$	
f (quark or lepton)	$f, \bar{f}^\dagger$	
$\bar{f}$ (Anti-quark or Anti-lepton)	$\bar{f}, f^\dagger$	
$\tilde{N}_i$ (Neutralino)	$\chi_i^0, \chi_i^{0\dagger}$	
$\tilde{C}_i^+$ (Chargino)	$\chi_i^+, \chi_i^{-\dagger}$	
$\tilde{C}_i^-$ (Anti-chargino)	$\chi_i^-, \chi_i^{+\dagger}$	
$\tilde{g}$ (gluino)	$\tilde{g}, g^\dagger$	

If we redefine the *theorem of spinor technology in paragraph iv)* [68], supposing that the total current from the spin fluxes to all particles appears to be integral to the soft photons (with uniformity of space-time at the quantum level), then the particle obtained with the application of a quantum transformation involving the Lorentz transformation given by $U(\Lambda)^{-1}\Psi(x)U(\Lambda)=D(\Lambda)\Psi(\Lambda^{-1}x)$, and substituting the terms $J^{\mu}(\xi)A_{\mu}(\xi)$ (reinterpreted in this theory as photon fluxes) for $\upsilon^{-1}\psi(p)$, which define a Euclidean transformation where υ is a matrix to be specified, replacing the kernel of (55), we have an equation that takes the form

$$
\begin{aligned}
W &= \frac{1}{2}\int \frac{(dp)}{(2\pi)^4}\psi_{\zeta_1\zeta_2}(-p)[\gamma^0(m-\gamma p)]_{\zeta_1\dot{\zeta}_1}\times \\
&\times[\gamma^0(m-\gamma p)]_{\zeta_2\dot{\zeta}_2}\psi_{\dot{\zeta}_1\dot{\zeta}_2}(-p)\left[\frac{1}{p^2+m^2-i\varepsilon}\right] \\
&= \frac{1}{2}\int \frac{(dp)}{(2\pi)^4}\mathrm{tr}[(-m-\gamma p)\Psi(-p)^{\mathrm{T}}\gamma^0(m-\gamma p)\Psi(p)\gamma^0]\times \\
&\times\left[\frac{1}{p^2+m^2-i\varepsilon}\right] \qquad (271)
\end{aligned}
$$

After this, we can use the properties of the symmetrical and anti-symmetrical matrices and engaging in some useful algebraic rearrangement, we arrive at the expression tr

$$
\begin{aligned}
&\mathrm{tr}[(-m-\gamma p)\Psi(-p)^{\mathrm{T}}\gamma^0(m-\gamma p)\Psi(p)\gamma^0]= \\
&-(p^2+m^2)\mathrm{tr}\Psi(-p)^{\mathrm{T}}\gamma^0\Psi(p)\gamma^0+m\times \\
&\times\mathrm{tr}(\Psi(-p)^{\mathrm{T}}\gamma^0[\gamma p,\Psi(p)\gamma^0])+\frac{1}{2}\mathrm{tr}([\gamma_p,\Psi(-p)^{\mathrm{T}}\gamma^0]\times \\
&\times[\gamma p,\Psi(p)\gamma^0] \qquad (272)
\end{aligned}
$$

where, for the two symmetries, we have

$$[\gamma p,\psi_a(p)\gamma^0]=\gamma^\mu i\gamma_5 p_\mu\psi_2(p)+\gamma_5 p_\mu\psi^\mu(p),(62)$$

$$[\gamma p,\psi_s(p)\gamma^0]=-\frac{1}{2}\sigma^{\mu\nu}i(p_\mu\psi_\nu(p)-p_\nu\psi_\mu(p))-$$
$$\gamma^\mu ip^\mu\psi_{\mu\nu}(p) \tag{273}$$

The evaluation is reduced to computing the traces of the matrices formed by multiplying linear combinations of Dirac matrices. Some of the algebraic properties of the Dirac matrices γ^μ imply that

$$\frac{1}{4}\mathrm{tr}\gamma_\mu\gamma_\nu=\frac{1}{4}\mathrm{tr}\gamma_\mu\gamma_5\gamma_\nu\gamma_5=-g_{\mu\nu} \tag{274}$$

while

$$\frac{1}{4}\mathrm{tr}\sigma_{\mu\nu}\sigma_{\kappa\lambda}=g_{\mu\kappa}g_{\nu\lambda}-g_{\mu\lambda}g_{\nu\kappa} \tag{275}$$

Likewise

$$\mathrm{tr}[(m+\gamma p)\psi_a(-p)\gamma^0(m-\gamma p)\psi_a(p)\gamma^0]=$$
$$-(p^2+m^2)[\psi_1(-p)\psi_1(p)+\psi_2(-p)\psi_2(p)+$$
$$\psi^\mu(-p)\psi_\mu(p)]+K(-p)K(p) \tag{276}$$

with

$$K(x)=2^{1/2}(m\psi_2(x)+\partial_\mu\psi^\mu(x)) \tag{277}$$

and

$$\mathrm{tr}[(-m-\gamma p)\psi_s(-p)\gamma^0(m-\gamma p)\psi(p)]=$$
$$-(p^2+m^2)[\psi^\mu(-p)\psi_\mu(p)+\frac{1}{2}\psi^{\mu\nu}(-p)\psi_{\mu\nu}(p)]+$$
$$J^\mu(-p)j_\mu(p)+\frac{1}{m^2}p_\mu J^\mu(-p)p_\nu J^\nu(p) \tag{278}$$

where

$$J^{\nu}(x) = 2^{1/2}(m\psi^{\nu}(x) + \partial_{\mu}\psi^{\mu\nu}(x)) \qquad (279)$$

K and *J* are structures of corresponding spinors for spin 0 and spin 1. We can observe the existence of additional terms that modify the photon sources in their most general form. This gives us an understanding of the production of photons in all its modalities!

The extension of photon sources by soft photon sources can be an elegant means of obtaining symmetrical frames of representation in terms of spinors in field theory. In this section, we only wish to establish this possibility by studying the infinitesimal transformations involved in the Lorentz transform and using the invariance of the metric (or Lorentz norm) in all quantum transformation processes that occur in the complex interaction of fermions and their phase spaces, possibly in the various steps of formation from the early universe up to the actual universe of today.

The interchanges of charges and changes of spin with these bosons establish, in quantum mechanics, the principles of creation and annihilation of all particles, from neutrinos to electrons. Electrodynamic methods have been useful in calculating the traces of particles and their energy states, as well as their meanings in space-time, since transitions in these quantum transformations produce trails that appear naturally in the calculations.

CHAPTER 5

SECOND RESULTS: HADRON PROLIFERATION AND EXTENDED PHOTON SOURCES

5.1. Second Results: Hadron Proliferation and Extended Photon Sources

Dark energy is gravitational energy with photon energy, filling all of space and and interacting with the matter-producing field effects of these bodies, which can be observed as maser and laser radiation. In fact, this step of the universe's development resulted in the proliferation of **H**-particles[29], which explains the quantity of hydrogen in the universe, being the fundamental chemical element in the composition of interstellar and sidereal bodies (stars, globular clusters, nebulae, and galaxies etc.).

In its initial state, with the interaction of large photon sources, the universe extended beyond the model of the de Sitter universe.

Theorem 2 (F. Bulnes). Dark energy lies beyond the fundamental interaction of baryogenesis and arises with the proliferation of **H**-particles in extended photon sources.

Theorem 3 (F. Bulnes). *Gravitational waves arise as energy waves produced by the interaction of dark energy with extended photon sources under **H**-torsion (brosons) in regions of the universe of major energy density.*

Proof. According to theorem 2 [9], if dark energy arises after baryogenesis and with proliferation of **H**-particles in extended photon sources, the torsion

---

[29] Hydrogen atoms.

remains. However, **H**-particles in extended photon sources are themselves the result of a set of heavy charged particles of photons with hadrons as the residual particles (particles that achieve additional processes and under conditions in which the hydrogen atoms make up the major material component of the universe[30] in our era). The relevant forces are strongly magnetic[31]. However, indicators of torsion are obtained when an interaction of an electromagnetic source with matter is realized. This has been proved experimentally using a Hall sensor conditioned to the movement of a magnetic dilaton in a trajectory of constant torsion whose matter agitates space and from which gravity emanates (from the kinematic tensor theory in duality with spinor waves [5]), interacting with the magnetic field of the proper dilaton (fig. 32).

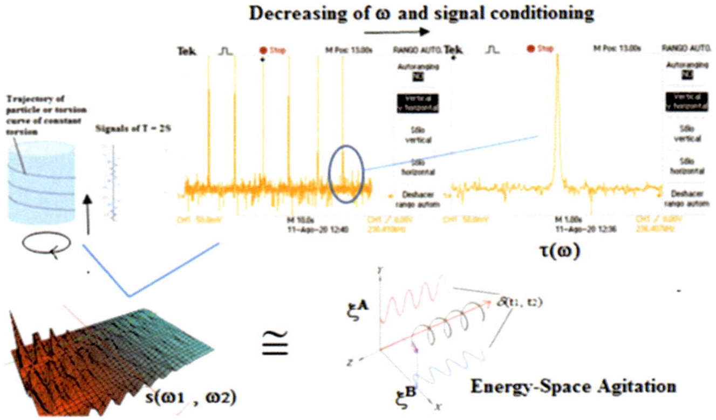

Fig. 32. Experiment with constant torsion.

As mentioned in chapter 2 [3], we can demonstrate that torsion is present as torsion waves, which are found in regions of major energy density, and as is the case for strong hadronic interactions (through the proliferation of **H**-particles), we can obtain gravitational waves that are measurable through Majorana fermion states ϒ (possibly using an advanced quantum Hall

---

[30] In fact, strong hadronic interactions emerge as residuals at this stage.

[31] This explains the intersidereal magnetic field observed up to today in several process forming interestellar bodies, for example, in the rotation and accretion of mature spiral galaxy arms.

sensor). Likewise, the table below presents an analogical comparison in the experiments to show torsion, gravitational waves, and the existence of dark energy in the universe, considering the electronics experiments and the possible spintronics experiments when use fermions.

Table 3: Analogies Between Experiments

Electronic Hall Sensor [3]	Property	Advanced Hall Sensor
Cylindrical Helix	Movement	$T_{ijk} = \varepsilon_{ijk}\dot{b}, \quad B^0 \propto \dot{b}$
$H = \frac{I}{2\pi}\frac{a^2}{1^3}$	Gauge Field	Axion Fields (for example Kalb-Ramond field)
$J^a = \sigma E^a$	Current	$J^{\mu}_{(A)} \equiv \psi\gamma_5\gamma^{\mu}\psi$
Torsion Energy	Spectral Density	H-Torsion
$\tau(\omega) = \frac{1}{2\pi}\left[\frac{(2.5V - \alpha)b}{l^3}\right] \times \frac{\sin(0.5\omega)}{0.5\omega}$	Evidence Torsion Signal	Gravitational Waves
U(1)	Transformation Group	SU(2), SU(2, 2), U(1)-anomalies
Electric Potential (= Volts)	Scalar Field	Kalb-Ramond field
$\tau(t_1, t_2) = \pm 2\pi \cot \Psi \times \mathfrak{F}\{K(t_1, t_2)\}\delta(t_1, t_2)$	Energy-Space Interaction	data $N_{\chi_1} - N_{\chi_2} = \cos 2\theta \propto n_{\psi}. > - < n_{\psi} >)$
----------	Dark Matter Detector	Hadrons Collider
CPT	Symmetry	CPTV
$R^{ab} = d\omega^{ab} + \omega^a_c \wedge \omega^{cb}$	Gravitational Field	$B^a = \in^{abcd} e_{b\lambda}(\partial_a e^{\lambda}_c + \Gamma^{\lambda}_{\alpha\mu} e^{\alpha}_c e^{\mu}_a)$

5.2. Some Post-Doctoral Results[32]

5.2.1. Some Universal Results for the Energy, Spectrum, and Generalization of Curvature

The study of the curvature at a general level, with well defined metrics and connections, begins with properties and intrinsic characteristics of the proper geometrical space to establish the invariance of the curvature of a space or manifold and determine that curvature is the most important geometrical characteristic that defines the shape of the space. From this, there arises a need to study curvature in more general geometrical spaces known as manifolds.

Likewise, if we consider space M as a Riemannian manifold, the Riemann

[32] Notes of Post-Doctoral Research Seminar Mexico-Ukraine-Russia

tensor $R_{\alpha\beta}$ can be expressed as the multi-linear functional

$$R_{\alpha\beta} = \sum_{\alpha,\beta} K_{\alpha\beta\mu\nu}(p)\mathbf{e}_{\alpha} \otimes \mathbf{e}_{\beta} \qquad (280)$$

$\forall p \in M$, where $K_{\alpha\beta\mu\nu}$ is the components of curvature as the tetra-linear mapping

$$K : V \times V \times V \times V \to F \qquad (281)$$

where F is a frequently the real set numerical field. In particular, if we consider M with a specified symmetry, then we have the Ricci tensor. This can be defined as a tensor made up of a set of components that describes part of the curvature of space-time. This is a symmetric second-order tensor, which we can define as

$$\mathrm{Ric}: T_p(M) \times T_p(M) \to \mathbb{R} \qquad (282)$$

with the correspondence rule

$$\mathrm{Ric} = R_{\alpha\beta} dx^{\alpha} \otimes dx^{\beta} \qquad (283)$$

The Ricci curvature is determined by the sectional curvatures of a Riemannian manifold. Likewise, these sectional or scalar curvatures can be determined by their principal curvatures. For example, if M is a cylinder with a radius equal to 1, then its principal curvatures are $k_1 = 0$ and $k_2 = -1$.

The sectional curvature is given by $S(\upsilon) = k_1 \cos\upsilon + k_2 \mathrm{sen}\upsilon = -\cos^2\upsilon.$

In general, the sectional curvature is given as $S(p) = \mathrm{tr}(r_p)$ where $r \in \mathrm{End}(T_p(M))$ (as a matrix of the curvature tensor whose trace is that of the sectional curvature). Then, transformed to a local coordinate system, this scalar curvature takes the form

$$S = \sum_{\beta=1}^{n} K'_{\beta\beta} = \sum_{\alpha,\beta} \varepsilon_\beta K_{\beta\alpha\alpha\beta} \qquad (284)$$

In particular, if $\mathrm{Ric} \in \mathrm{symm}\, T_p(M)$ and if $\forall \xi \in T_p(M)$ and $\xi = |\xi| \delta_n$, then the sectional curvature along n-directions of the basis $\{\delta_1, \ldots, \delta_n\}$, can be written as

$$S(p) = \mathrm{Ric}(\delta_\beta(p)) = \sum_{\alpha \neq \beta} K(\delta_\alpha \wedge \delta_\beta)(p), \qquad (285)$$

If $n = 2$, then

$$\mathrm{Ric}(u) = K_G(p)(|\xi^1|^2 + |\xi^2|^2) \qquad (286)$$

where $K_G(p)$ is the Gaussian curvature of M in $p \in M$.

In terms of the principal curvatures for an n-dimensional symmetric manifold M, this Gaussian curvature can be written as

$$K_G(p) = k_1 k_2 \cdots k_n \qquad (287)$$

Furthermore, its mean curvature takes the form

$$H(p) = [1/(n-1)](k_1(p) + \ldots + k_n(p)).$$

In particular, with an orthonormal reference frame of $\{\mathbf{e}_1, \mathbf{e}_2\}$, then, relative to this frame, the unique components of the curvature are

$$K_{1212} = K_{2121} = -K_{1221} = -K_{2112} = K$$

and for the Ricci tensor

$$K'_{11} = K'_{22} = K \text{ and } K'_{12} = 0$$

which is to say that the curvature of all symmetrical spaces, such as spheres, cylinders, and one-sheet paraboloids etc., result in the matrix of their Riemann tensor, with non-null diagonal components and all others equal to zero.

The sectional curvature in the 2-dimensional case will simply be

$$S(p) = 2K_G(p), \quad \forall p \in M$$

Between the Ricci, Gaussian, and sectional curvatures, we find an interesting relation known as the Bishop inequality

$$\text{Ric}(\xi, \xi) \geq \kappa(n-1)|\xi|^2, \forall \xi \in T_p(M)$$

with $n = \dim M$ and where κ is the constant sectional curvature of model spaces M_k, which has a simple locally isometric connection to *M*.

For example, spaces such as spheres are locally isometric model spaces in relation to all simply connected convex complete manifolds. This is a consequence of the existence of a diffeomorphism between *M* (with such characteristics) and the *n*-dimensional sphere S^n, called the Gauss mapping of *M*, which satisfies $\text{Vol}M \leq (\text{Vol}(S^n))/\kappa^{n/2}$ where, if *M* is isometric to S^n, then the last inequality is reduced to $\text{Vol}M = (\text{Vol}(S^n))/\kappa^{n/2}$.

Considering a change of metric *g* due to multiplication by a conformal factor, the conformally-related metric is given as $\hat{g} = e^{2f} g$ and the Ricci tensor is given by

$$\overline{\text{Ric}} = \text{Ric} + (2-n)[\nabla df - df \otimes df] + [\Delta f - (n-2)\|df\|^2]g, (288)$$

where the differential operator $\Delta = d * d$ is the (positive spectrum) Hodge Laplacian. That is to say, the opposite of the usual trace of the Hessian.

Likewise, considering our Riemannian manifold M (which could even be a pseudo-Riemannian manifold) and some of general relativity, the trace-free tensor of our manifold is the tensor defined by

$$Z_{\alpha\beta} = \mathrm{Ric} - \frac{S}{n} g_{\alpha\beta} \qquad (289)$$

where Ric is the Ricci tensor; S is the sectional or scalar curvature; g is the metric tensor; and n is the dimension of M. Then, the trace of the tensor $Z_{\alpha\beta}$ is automatically annulled, i.e. $Z_{\alpha\beta} g^{\alpha\beta} = 0.$

If $n \geq 3,$ then the trace-free Ricci tensor vanishes if, and only if, Ric accords to some constant $\lambda,$ characterizing the pseudo-Riemannian manifold as an Einstein manifold. These are the spaces studied in relativity.

As an application of the curvature concept in such manifolds, and due to general and special relativity, the importance of obtaining a curvature for the cosmos is primordial to the understanding of the geometrical scenarios that produce a gravitational field solely for the presence of matter. The curvature usually appears as an observable of the gravitational field, which is measured by the evaluation of the field tensor $R_{\alpha\beta}$ of the Einstein equations on the geodesics of the universe.

Due to the fact that the universe can be seen as an intrinsic space the actions of which produce proper movements, which conserve invariants of all the elements of the proper universe[33], it is feasible to model the cosmos as a homogeneous space. The equivalence of classes or orbits concerns the reductive homogeneous spaces of the cosmos with the goal of calculating observable aspects of regions the volume of which can be calculated through certain integral operators on the orbits or classes of said regions (reductive homogeneous spaces).

---

[33] That is to say, the universe seems to be of the same shape and with the same properties in any region of itself.

Likewise, if *M* is our cosmos, the action on the homogeneous space for the *G*-structure defined for the quotient group subjacent in the differentiable structure of *M* induces actions on the vector bundles of *M*.

This allows us to define invariant connections in reductive homogeneous spaces and determine the value of the defined functions on sections of the *G*-invariant vector bundle, defining observables of the corresponding endomorphisms to the vector fields of the vector bundle sections. These observables geometrically and physically characterize and define the intrinsic form of the cosmos.

One of the advantages of characterizing space *M* (the universe) as a homogeneous space is the determining of any observable of proper space (volume, area, density, curvature, mass, torsion, and magnetic and electric permeability etc.) through the contribution of these observables in a space-time region topologically identified as having closed submanifolds.

These regions of the universe have the same structure as the total cosmos and their transformations produce proper movements in the submanifolds. Then we can construct a regular *G*-invariant measure on the submanifolds of certain arbitrary codimensions and permit the evaluation of the invariant integral operators used to determine the value of the observable.

In this way, "the whole is known through its parts", with a relation between the parts accepting a transitive action of the subjacent semi-simple group in the manifold *M* and with trivial action on the proper universe *M* from these submanifolds[34].

Likewise, we can use a version of the Riemannian curvature for temperate distribution on differentiable manifolds, which is more general than that belonging to an cohomological isomorphic integral operator, related in a certain sense to the de Rham cohomology with topological vector spaces of infinite dimension and that generalizes in the classic sense.

Thus, we consider the value of integrals in closed classes of homogeneous space *G/H*, where *H* is a closed subgroup of *G* corresponding to the

[34] Theorem on the generalization of curvature as tempered distribution [69-71].

differentiable manifold *M*. We determine the integrals on the homogeneous space *G/H* of the form

$$\int_{G/H} f(x)dx = \int_{G/H} f(x\sigma)dx_{\sigma} \tag{290}$$

where $f \in C_c^{\infty}(M)$.

In particular, these integrals take the form of the Radon transform on generalized spheres

$$\mathcal{R}(f) = \int_{\Xi} f(x)d\sigma = \int_{g\Xi} f(x)d\sigma_g \tag{291}$$

If we consider a *K*-invariant *G*-structure given by the space $S_G(M)$ on a reductive homogeneous space $M \cong G/H_M$, given by the quotient space G/G_0 with decomposing $\mathfrak{t} = \mathfrak{h} + \mathfrak{m}$, then the cosets of *K*-invariant connections in $S_G(M)$ isomorphically correspond to the linear mappings of differential operators $\widetilde{\nabla}_{\mathfrak{m}} : \mathfrak{m} \to \mathfrak{g}$.

These are particularly consigned to smooth submanifolds (cycles) Σ smoothly embedded in *M* as the connections

$$\nabla : T_p(D'(M)) \to T_p(D'(\Sigma)) \tag{292}$$

Thus, the Radon transform in the closed submanifold context is a tempered distribution with images or co-cycles $\mathcal{R}_{\sigma}(\Sigma)$ where σ is the smooth embedding of Σ in $D'(M)$. Their extrinsic curvature is obtained as the integral (another way to write the sectional curvature with the idea of Riemannian curvature)

$$K = \mathfrak{hess} \int_{\Sigma^y} f(x)d\mu_y(x) \tag{293}$$

which conforms to a theorem that I worked on from 1998 in an inverse problems seminar at UNAM and published in year 2001, also at UNAM [18]. Later, I published some results on its applications in astrophysics and signal analysis for measuring the curvature of a 2-dimensional manifold model of the universe with hyperbolic structures via horocycles. The results were published in IBEROMET VII [62] (Cancun Quitana Roo, Mexico) in 2002. This was considered an important result in aerospatial technology. All results up to 2006 were presented in the book "Research on the Curvature of Homogeneous Spaces" (in Spanish "Investigación sobre Curvatura de Espacios Homogéneos") [61], published by the Research Department of Mathematics and Engineering, TESCHA.

However, the calculation of integrals on cycles and co-cycles of a Riemannian manifold to determine their curvature can be generalized to the context of complex differentiable manifolds, even involving singularities. Studying some unitary representations of $SL(2,\mathbb{C})$ and $SU(2, 2)$, we can obtain results on the curvature from which we can determine a general algorithm through an integral operator for L^2-cohomology. This can be generalized by an integral operator cohomology that preserves the complex inner product determined for the integrals of the fundamental 2-forms. The product is the curvature value in a more general context. It also incorporates twistor geometry for the treatment of singularities through the homogeneous vector bundles. These bundles interpret the singularities as zeros of polynomials of rings of the corresponding algebra considered inside the structure of a complex Riemannian manifold as $\mathbb{M}$.

Using Big Bang theory, the concept of the universe's expansion, black holes, and super-string theory we can establish regions of space-time with a surface tending to zero (and a volume tending to zero) and thus with infinite curvature.

Einstein's model without the cosmological constant supposes that the universe is experiencing indefinite expansion with curvature that is infinitely tending to zero. That is to say, this is a model of an inflationary universe that is asymptotically converted in a plane. Some indications of the behavior of light, the "gauge" invariance of the Maxwell equations under inertial translations, the observation of certain symmetries and super-symmetries in some strong interactions, and the hyperbolic nature of the non-symmetrical field equations presuppose a hyperbolic structure of the universe and a hyperbolic model of the cosmos. This helps us construct the connection between them and, using the Riemannian metric of hyperbolic model space, we can determine the curvature of the cosmos. However, some interactions of quantum type must be taken into consideration in the study of space-time singularities, since these provoke space-time deformations that surround the singularities. Such deformations can be measured through a curvature that must be constructed for cohomological study, establishing equivalences between deformed spaces and their dualities in the microstructure of space-time.

The complex cohomologies by which we can establish certain hypotheses on the cycles and the line homogeneous bundles are of great value. They allow us to develop an approach with $(n-1)-\bar{\partial}$-cohomology, which can be used for handling the infinite and indeterminate aspects of curvatures in spaces with singularities. As such, can we generalize this idea to obtain the curvature of any curved space?

For this purpose, we must reformulate the scheme of the Radon transform on cycles in M through the use of line bundles involving certain projective spaces. Using the preliminary ideas of John and Penrose, the integrals below have a form that can suggest the construction of functions in cosets of modulo functions defined in hyperspaces (an example being hypercomplex functions).

$$\phi(w, x, y, z) = \int_{-\infty}^{+\infty} f(w + x\zeta, y + z\zeta, \zeta) \tag{294}$$

or

$$\phi(w,x,y,z)=\int_{S^1} f((w+x)+(y+iz))\zeta(y-iz)+(w-x)\zeta,\zeta)d\zeta$$ [35]

Thus, twistor theory can also be utilized, being the first indication that certain twistor transforms[36] (the Penrose transform) establish an isomorphism between the space of the first $\bar{\partial}$-cohomology in relation to positive twistors and the space of holomorphic solutions of the corresponding massless field equations in a future tube. This problem of cohomological classes firstly indicates the construction of an integral operator cohomology in relation to the classification of differential equations due to the integral operators that solve such equations. The problem of handling infinites aligns more with the geometric handling of a function space of analytic type. Likewise, only an intersectional cohomology on geometrical invariants is necessary to establish a way of handling the infinites through the image of these singular points, which do not depend on their position in space. In addition, the topological invariant in this case results in a proper intersectional cohomology. Some exercises in this respect have been developed in relation to twistor theories to handle the infinites of the curvature tensor of space-time and the interpretation of particles in the frame of relativity and quantum phenomena, which could be interpreted as singularities or space-time distortions in the macroscopic frame.

In this context, other integral transforms arise, from which we can generalize, in a particular sense, the exploration of the geometrical Radon transform, but applied to the fields of QFT and QED.

This study requires us to use a generalized version of a $\bar{\partial}$-cohomology with complex Riemannian manifolds and invariant integral operators on the manifold cycles. This new formulation of a $\bar{\partial}$-cohomology (which can be formulated through their extension via field ramifications) of invariants in cycles establishes an integral operator cohomology as a basis for the theory

[35] Whose geometrical interpretation is as a Radon transform on the lines of a flag manifold $F=\{L|L\subset\mathbb{R}^4\}$.

[36] The Penrose transform is a twistor transform in a certain case, context, or dimension of mapped spaces.

of curvature. As such, a conjecture that seems natural in terms of generalizing the curvature when handling infinites in space-time and, in general, operates on differentiable manifolds with singularities, allows us to devise an integral operator model of curvature belonging to a cohomological class that is admissible in space-time and can explain the general tendency of the Weyl tensor (anti-symmetric part of the Riemann tensor) to expand with time along the geodesics inside the future cone of their causal structure through the kernel of an integral operator.

Thus, we can propose the following complete cosmological research scheme for evaluating the integrals, which must be realized on the orbital submanifolds corresponding to the conformal structure of the complex Riemannian manifold $\mathbb{M}$. We will use this as a model, along with the deformed twistor space previously modeled to be applicable to the situation in question. Here, we should note that this can be obtained from a geometry derived for the study of the deformation of the categories in deformation theory, which relate the path integrals used in string theory to the twistor theory in singularities.

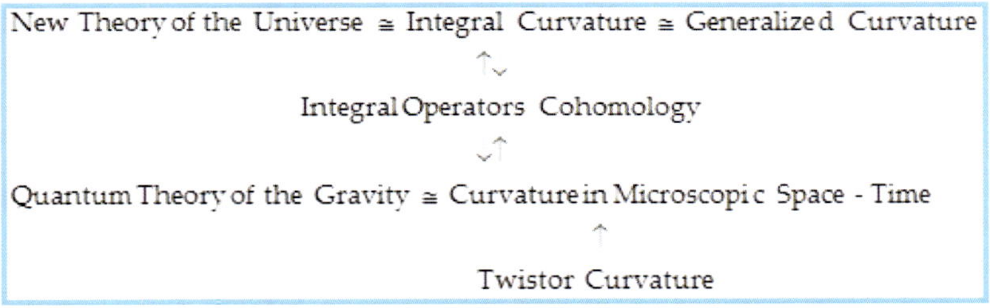

Fig. 33. Complete cosmological research scheme from the integral geometry point of view.

The Penrose transform can be combined with the Feynman path integrals to establish a particular hypothesis on the study of contour deformation. Then, considering a good *D*-module theory that gives results for deformation theory [72], in the study of geometrical ramifications in field theory and other studies on the resolution of the field equations, we consider the Penrose transform a good apparatus from which to obtain and generate solutions through kernels for the different differential operators (flat and curved operators etc.).

From this, we have developed studies in field theory, designing a microscopic homogeneous space $G[[z]]/\mathbf{X}$ for the study of quantum gravity. We can create a quantum gravity sensor based on the advanced concept of curvature energy that I obtained and developed in the study of electronic sensors [10, 11, 17, 25]. Here, at this point, we have to emphasize the use of the generalization of the curvature obtained in 2001, but translated in terms of signals and systems analysis considering Gaussian curvature, referring to the principal curvatures mentioned before in spectral form. Here, the co-cyles are energy pulses with geometrical invariance and the curvature of which is obtained through analysis of their curvature energy constructed by co-cycles of the corresponding integrals

$$\kappa(\omega_1, \omega_2) = \int_M \kappa(p, \phi) e^{-j(\omega_1 t_1 + \omega_2 t_2)} dp d\phi \qquad (295)$$

For the quantum context, we can obtain regular representations with realizations on the $L^2(G/B)$-cohomology of integral operators as an $L^2(\mathfrak{g}/\mathfrak{h})$-cohomology of integral operators [69, 70, 73]

$$< \Lambda_{\tilde{g}}(f), f >= \bigoplus_{\zeta \in K^{\wedge}} < \Lambda_{\tilde{g}}(V(\zeta)), f > \qquad (296)$$

Better representations are realized as Penrose transforms when G is a Lie group of infinite dimensions (non-compact) and H is locally compact.

The focus of my research, over many years of studying specialized curvatures, has been to generalize and extend the concept of integral curvature obtained through the Radon transform in the primary context of differentiable manifolds, such as discontinuous manifolds, projective manifolds, vector bundles, Lie groups, and Lie algebras, etc. Furthermore, along with this generalization we wish to unify the concept of curvature, studied in a different way for differential and integral geometry. Their unification will be obtained through the generalization of the concept of curvature on any differentiable manifold class. Also of interest to us, and this is a central concept of notes on postdoctoral research presented in a number of different seminars, is the interpretation of curvature through energy. This should be considered in the light of recent results concerning the electronic application of sensors and transducers, combined with important publications giving the results obtained for the curvature spectra $\kappa(\omega_1, \omega_2)$ and using diffeomorphisms, through which we have obtained the spectrum of curvature and developed a corresponding scheme for the Spec functor. Here, our focus is on field theory developed in diverse mathematics papers and using integral geometry for the study of the deformation of derived categories.

5.2.2. Conformality in the Curvature Spectrum of Space H

H is the space of the energy states that determine the Hamiltonian densities of the energy spectra due to curvature. Additionally, these densities can be superposed to integrate a curvature spectrum, as established in [74] (see figures below).

From these considerations, the question of how we can recognize the spin 2-particles or gravitons, which determine the gravitational energy scenario useful in field theory, to generate curvature spectra directly from the action of the gravitational field on space arises. The generalized stress tensor conservation law given by the equation

$$\nabla_{\mu} t^{\mu\nu}(x) = 0 \qquad (297)$$

will fail inside particle sources (such as those shown in fig. 3), unless one recognizes the pre-existence of the energy and momentum that is transferred to the emitted particle. However, there is not an electromagnetic analogue to the graviton source problem. The photons are electrically neutral, whereas the gravitons carry energy-momentum, which also must be transferred rather than created within the source.

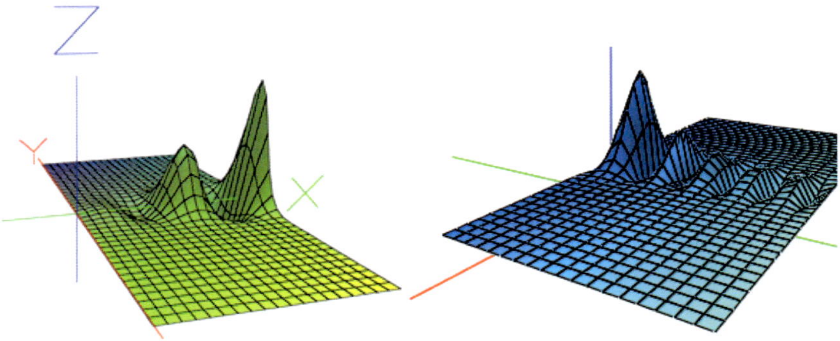

Fig. 34. Superposing of the Hamiltonian state densities measuring curvature. Using the formal invariance under conformal transformation, the **H**-states can be determine not only for the first curvature, but also for the second curvature. These energy states arise in a natural way during the study of fermion states interacting with the gravitational background and we designate a gauge field as a "dilaton" of magnetic nature, which can evade the strong radiation that arises in this interaction (see experiments and theory in chapter 2). The study of strong gravitational sources realized from the point of view of the kinematic tensor $A_{\mu\nu}$ in relation to the stress-matter tensor offers a method to determine the existence of space-time singularities (see fig. 35 B)).

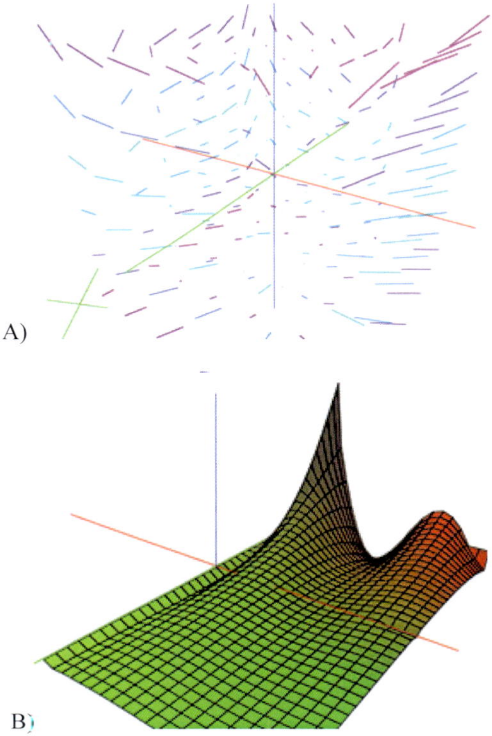

Fig. 35. A) Gravitational field of perturbed spherical symmetry. B) Cylindrical gravitational waves determining a gravitational source in space-time.

Thus, we introduce a source that can provide the correct gauge transformation behaviour of charged particle sources. However, these source problems can be viewed in terms of the explicit search for $g_{\mu\nu}$-dependence, which gives various sources with the correct response for general coordinate transformations. Considering all that has been stated previously, the simplest example is a scalar source $\Omega(x)$ appearing in the action integral through the term $\int (dx)(-g(x))^{1/2}\,\Omega(x)\phi(x)$. However, these terms are useful for establishing graviton emission through certain special functions, such as impulse functions, whereas the terms

$\int (dx)(-g(x))^{1/2}\Omega(x)\phi(x)$ takes the form in the frame of the stress tensors $t_{\mu\nu}$ for high radiation emission

$$\Xi_{\lambda} = \int (dx)(1+h)\phi(\Omega(x) + X^{\mu}(h)\partial_{\mu}\Omega) \qquad (298)$$

where h is the densities defined in

$$\mathcal{R}_{\gamma}^{-1}\mathrm{Hom}_{\mathrm{K}}(\mathrm{Ad}(G^{0})\otimes D'(M),\, C^{\infty}(\Sigma))\mathcal{R}_{\gamma} = \mathcal{R}_{\gamma}(\mathrm{C}^{\infty}(\Sigma)),$$ [69]

Then, assuming that the graviton detection sources do not overlap the Ω-support region, one can use the source-free weak gravitational field equations and derive

$$\partial_{\lambda}X^{\lambda} = \int (dx')(dx'')f^{\mu}(x-x')f^{\nu}(x'-x'')\partial_{\mu}^{''}\partial_{\nu}^{''}h(x'') \qquad (299)$$

where $f^{\mu}(x-x')$ is one of the familiar class of functions, such that

$$\partial_{\mu}f^{\mu}(x-x') = \delta(x-x') \qquad (300)$$

as mentioned previously.

Additional work in the term inside integral (298) $\phi(\Omega(x) + X^{\mu}(h)\partial_{\mu}$ produces local diffeomerphism, creating field equations in this context. These can be transformed into the Hill equation of the form

$$\nabla^{2}\phi + \Phi\,\phi = 0 \qquad (301)$$

In the study of integral geometry, the *G*-structure of the complex Riemannian manifold, furthermore of the *K*-invariance of their submanifolds or cycles under the integral transforms, takes the form of Bulnes's equation [1, 9]

$$\mathcal{R}_{\sigma}(\mathcal{R}_{\sigma} \otimes \mathrm{Ad}) = 0 \qquad (302)$$

Both equations present generalizations of the harmonic oscillator and their curvature solutions will result in a spectrum of curvature. Equation (302) results in the incorporation of the intrinsic image of curvature of the submanifold, smoothly embbeded in the space $D'(M)$.

The considerations realized before have as their objective the establishment of the curvature spectrum. As a result, we obtain the incorporation of general relativity into quantum field theory so as to obtain the spectra of the curvature tensor $R_{\mu\nu}$ through other tensors that are directly or indirectly related to this curvature tensor and that are a product of the diffeomorphism applied to the microscopic structure of homogeneous space of form G/G_0.

In addition, the conformality in the spectrum of curvature of space can be established as a condition for measuring the duality of the curvature. This appears as a pseudo-symmetrical condition through which we can generalize the spherical mapping of Gaussian curvature, but, in this case, with the energy of Hamiltonian space in the gravitational field source (see fig. 36).

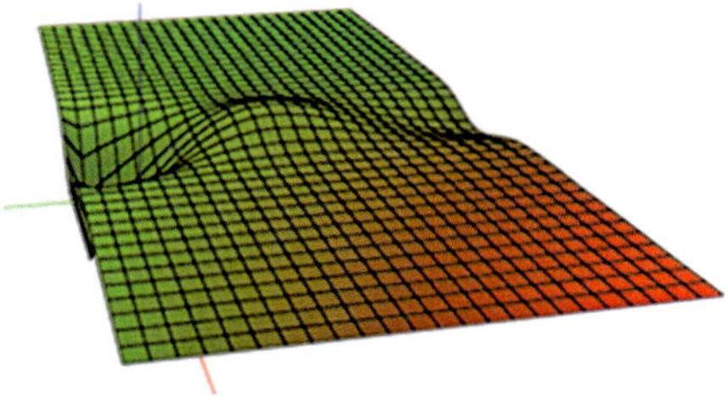

Fig. 36. Appearance of the gravitational field source. This can correspond to a 2-D surface model of a source as detected in a galaxy located in the Centaurus A, Ngc 5128 constellation (fig. 37).

Fig. 37. Galaxy in the Centaurus A, Ngc 5128 constellation.

5.2.3. Dark Matter and its Relation to Particle Physics: Some Evidence of Torsion Through Expansion

We consider again the Einstein field equation

$$R_{\mu\nu} - \frac{R}{2} g_{\mu\nu} = \xi T_{\mu\nu} \qquad (303)$$

with the stress-energy-matter tensor expresion

$$T_{\mu\nu} = (\rho + p) u_{\mu} u_{\nu} + p g_{\mu\nu} \qquad (304)$$

Here, the terms for matter density ρ and pressure p are considered apparent in the Einstein universe in relation to homogeneitity and isotropy. Likewise, in Einstein's space-time model, the geometry, as defined by the Minkowski metric, takes the following form according to the transformation by spherical coordinates

$$ds^2 = -c^2 dt + a^2(t)[dr^2 + r^2(d\theta^2 + \sin^2\theta d\varphi^2)] \quad (305)$$

We can deduce the Friedman equation with inflation H

$$\left(\frac{\dot{a}}{a}\right) = H^2 \quad (306)$$

where H is

$$H = \pm\sqrt{\frac{\xi}{3}\rho} \quad (307)$$

Thus, we can deduce the equation of the conservation of matter

$$\dot{\rho} + 3H(\rho + p) = 0 \quad (308)$$

However, taking into account the experimental measurements and observed facts, and considering the luminosity of sidereal objects in the universe, which are accelerating away along with others in the negative solution in (307), we can establish

$$d_{\text{Luminosity}}(z) = \frac{c}{H}\left[z + \frac{1}{2}(1 - q_0)z^2 + \ldots\right] \quad (309)$$

resulting in a Taylor expansion for the distance between two arbitrarily denoted sidereal objects (*one is the "observer" with coordinate O, while the other is located at point z*). In fact, through this asymptotic development we can obtain a good approximation of the localization of sidereal objects in Einstein's universe with expansion connected to negative pressure, giving it a negative value. In addition, we have

$$q_0 = -\frac{\ddot{a}\,a}{\dot{a}^2} = -\frac{\ddot{a}}{a}\frac{1}{H^2} < 0 \quad (310)$$

resulting in acceleration if the pressure is negative.

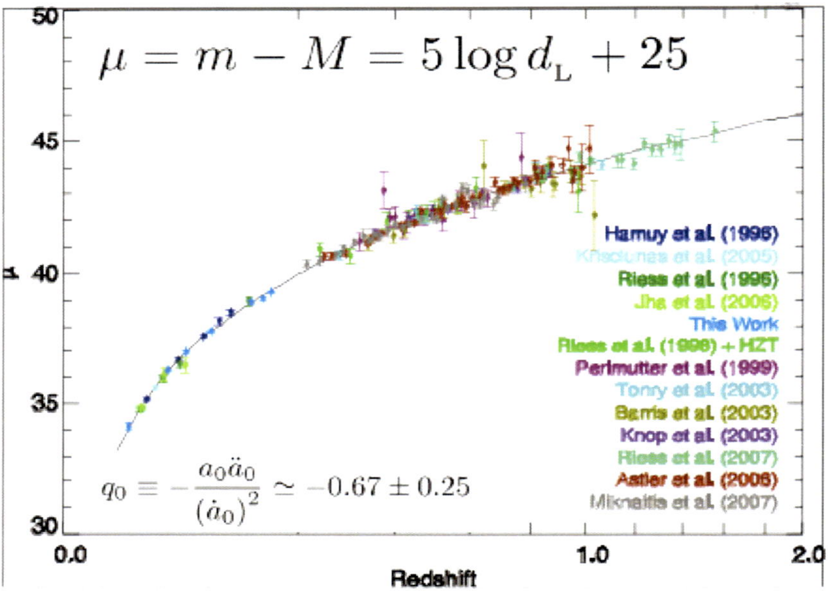

Fig. 38. Sequence of measurements taken over many years (1995-2007) giving observational data on expansion due to acceleration when the pressure is negative.

At greater precision, we have

$$\frac{\ddot{a}}{a} = -\frac{\xi}{6}(\rho + 3p) \qquad (311)$$

from which we infer

$$p = -\frac{\rho}{3} \qquad (312)$$

However, there is very little matter to expand the universe. The mass residuals are very sporadic (with great exapnses of space and scarcely one hydrogen atom per *cubic parsec*) and with little concentration of matter (see fig. 39). Furthermore, such concentrations are more likely to be perceived when analyzing the redshift.

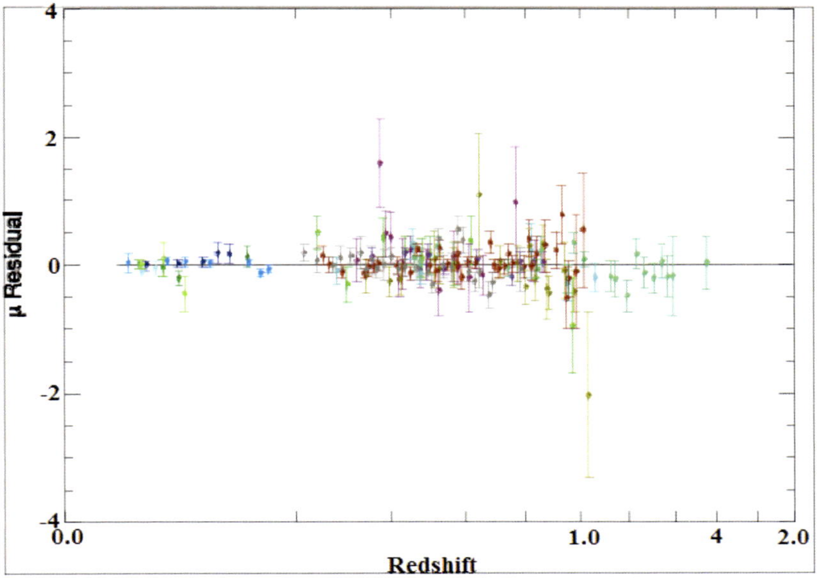

Fig. 39. Residual mass is very sporadic (with expanses of space that scarcely have one hydrogen atom per cubic parsec) and with little concentration of matter.

The question remains as to what is causing the acceleration. Einstein presented a number of treatises in which the cosmological constant was understood as a fluid stress tensor given by

$$T_{\mu\nu} = -\frac{\Lambda}{\xi} g_{\mu\nu} \qquad (313)$$

The reason for this was

$$\nabla^{\mu}(\Lambda g_{\mu\nu}) = \Lambda \nabla^{\mu} g_{\mu\nu} = 0 \qquad (314)$$

The observed facts have established that expansion is due to the acceleration of sidereal objects. However, with density and preasure in the regime of the cosmological constant, the energy-matter tensor (313) is given as

$$p_\Lambda = \frac{\Lambda}{\xi}, \qquad \rho_\Lambda = -\frac{\Lambda}{\xi} \qquad (315)$$

which could implicate

$$\omega_\Lambda = \frac{p_\Lambda}{\rho_\Lambda} = -1 \qquad (316)$$

suggesting (308). Rewriting equation (306) in terms of the cosmological constant suggests that it can cause acceleration. That is to say

$$\left(\frac{\dot{a}}{a}\right)^2 = \frac{\xi}{3}\rho = \frac{\Lambda}{3} = cte \qquad (317)$$

Resolving the differential equation (317) with respect to time, we obtain

$$a(t) = e^{Ht} \qquad (318)$$

which implies

$$\frac{\ddot{a}}{a} = H^2 > 0 \qquad (319)$$

This is obvious since the pressure is negative and, thus, there is expansion. One of the space-time models that can be adjusted to account for the reality of the cosmological constant (a space-time model when $\Lambda = -1$), is that of de Sitter spacetime (as for inflation) (see fig. 40).

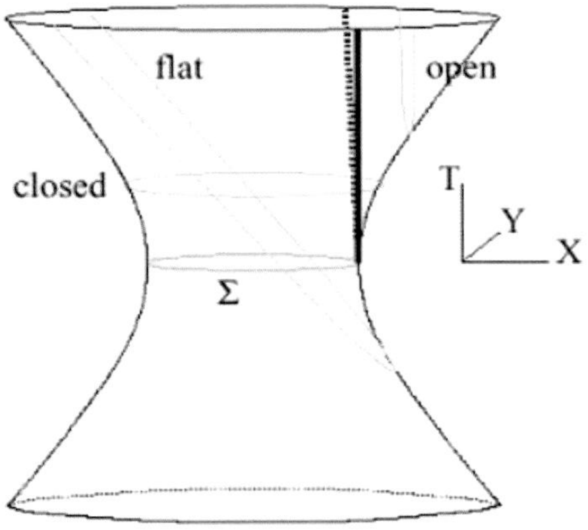

Fig. 40. Space-time model when $\Lambda=-1$.

Is it necessary to change the scale of observations to establish the real causes in deep space-time of expansion without the cosmological constant?

The answer is yes, the fundamental scale of matter density must be modified according to the energy-matter density at the microscopic scale to understand the low existence of matter in such a large volume of space. For example, observations have been made at this scale of the extragalactic universe (see fig. 41), considering a matter density of the order $\rho_\Lambda \approx 10^{-45} GeV \approx \rho_{cri} \approx (10^{-3} eV)^4$ and with microscopic length $L_\Lambda \approx 10^{-5} m.$

Thus, with these new considerations at the microscale, we can establish the origin of acceleration and therefore the universe's expansion, considering the study presented in the results of section 5.1 of this chapter. We may add [9], in which the proliferation of dark energy is realized when there is a greater proliferation of fermions, with H-particles in extended photon sources. This is the majority produced accounting for up to 70 % of the fundamental matter types (see fig. 42).

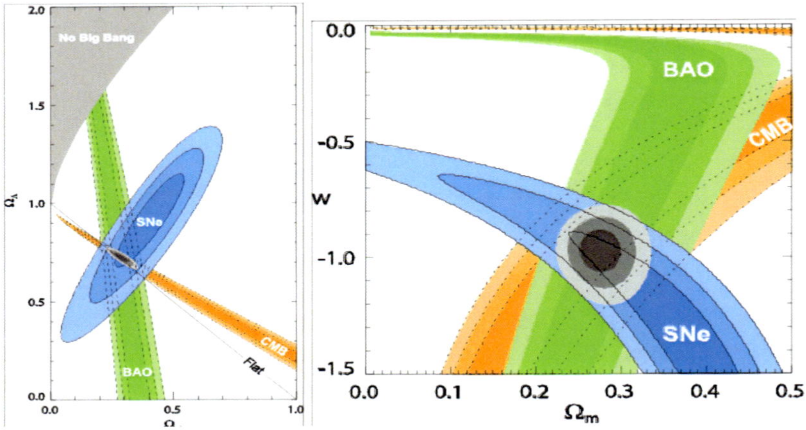

Fig. 41. Two mappings with different microscales of density derived from the cosmological constant.

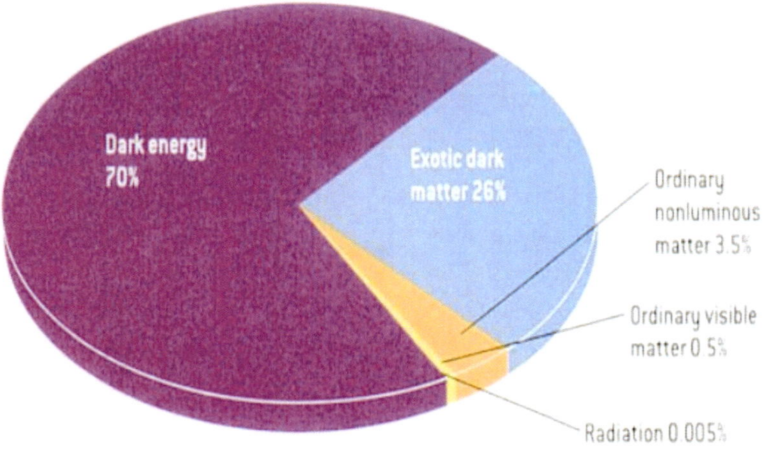

Fig. 42. Composition of matter types. Dark matter is a consequence of dark energy with the proliferation of hadrons.

However, the weight of the vacumm and the phase transtions in the early universe suggest other factors have contributioned[37] to the acceleration of the universe's expansion. For example, at the microscopic level, considering Fock space (as a box with microscopic dimensions (see fig. 43)), the evolution

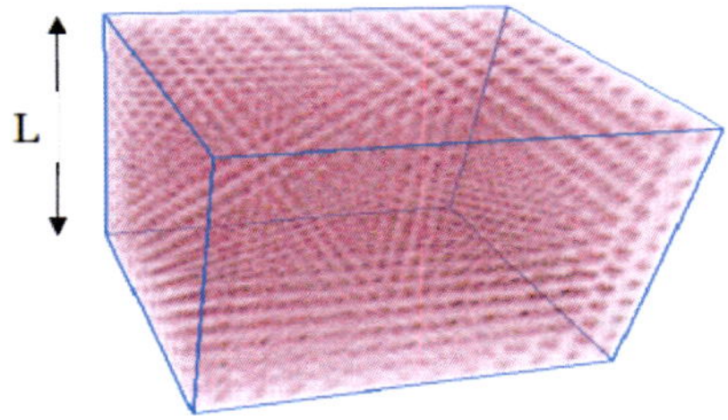

Fig. 43. Contribution to the vaccum of a simple quantum box. A scalar field in a finite box.

is given by

$$\Phi(t,\mathbf{x}) = \sum_{\mathbf{k}} \frac{1}{\sqrt{2L^3\omega_{\mathbf{k}}}}\left(a_{\mathbf{k}} e^{i\mathbf{kx}-i\omega t} + a_{\mathbf{k}}^{\dagger} e^{i-\mathbf{kx}+i\omega t}\right) \quad (320)$$

in which the creation and anhihilation operators satisfy the orthogonal relation

$$[a_{\mathbf{k}}, a_{\mathbf{p}}^{\dagger}] = \delta_{\mathbf{kp}} \quad (321)$$

and having the inflation

$$H = \sum_{\mathbf{k}} (a_{\mathbf{k}}^{\dagger} a_{\mathbf{k}} + \frac{1}{2})\omega_{\mathbf{k}} \quad (322)$$

---

37 $\Lambda_{eff} = \Lambda + \xi\langle\rho\rangle_{vev} + \xi\langle\rho\rangle_{PT} + \ldots$

where the product of the operators $a_{\mathbf{k}}^{\dagger}a_{\mathbf{k}}$ determine the particle number in the box. For (322), we have

$$\omega_{\mathbf{k}} = \sqrt{\mathbf{k}^2 + m^2}, \qquad k_j = \frac{2\pi}{L} n_j \tag{323}$$

and so, the vaccum contribution is given by

$$\langle 0|H|0\rangle = \frac{1}{2}\sum_{\mathbf{k}}\omega_{\mathbf{k}} \rightarrow \frac{1}{2}\left(\frac{L}{2\pi}\right)^3 \int \omega(k) d^3\mathbf{k} \tag{324}$$

for which we have the state of matter

$$\langle \rho \rangle = \frac{1}{2}\frac{1}{(2\pi)^3}\int d^3\mathbf{k}\sqrt{\mathbf{k}^2 + m^2} = \frac{M_c^4}{16\pi^2}\left(1 + \frac{m^2}{M_c^2} + \dots\right) \approx 10^{122}\rho_{cri}, \tag{325}$$

obtained with $M_c = M_{PI}$.

Another contribution comes from the electroweak phase transition, defined through the electrical potential

$$V = V_0 + \frac{1}{2}m^2\phi^2 + \lambda\phi^4 \tag{326}$$

where the comparison with the critical energy density is given by

$$\langle \rho \rangle_{PT} = V_0 - \frac{m^4}{4\lambda} = V_0 - \frac{\lambda}{4}(246GeV)^4 \tag{327}$$

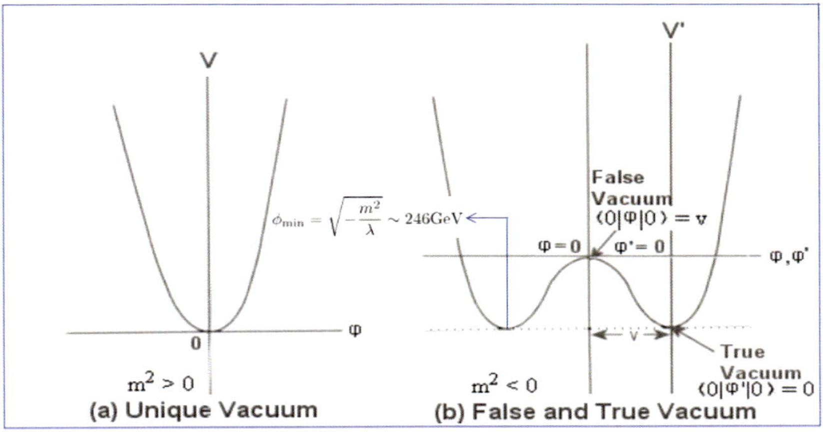

Fig. 44. Enormous in comparison with the critical energy density.

Thus, the cosmological constant problem persists according to the following assertions:

i) In general relativity, the vacuum gravitates.

ii) The vacuum energy density is made up of huge disconnected pieces, while the observed value is tiny, as mentioned in footnote 37. In Λ_{eff}, all the contributing components are huge, while the results for Λ_{eff} are very small.

How then can we establish an appropriate landscape for inflation+string theory?

We could establish an energy functional to consider supersymmetry as symmetry that relates bosons to fermions. The simplest example of this is the Wess-Zumino model given by the Lagrangian

$$\Lambda_{eff}, \eta_{\mu\nu}(\partial^{\mu}\Phi)^{\dagger}(\partial^{\nu}\Phi) + i\overline{\Psi}_{\dot{\alpha}}(\overline{\sigma}^{\mu})^{\dot{\alpha}\alpha}\partial_{\mu}\Psi_{\alpha} + F^{\dagger}F, (327)$$

This Lagrangian is invariant under the following transformation characterized by ζ where, from (327), we have

$$\delta\Phi = \sqrt{2}\zeta^{\alpha}\Psi_{\alpha} \qquad (328)$$

Here, we obtain a fermion from a boson. Furthermore, we have

$$\delta\Psi_{\alpha} = -i\sqrt{2}\left(\sigma^{\mu}\right)_{\alpha\dot{\beta}}\overline{\zeta}^{\dot{\beta}}\partial_{\mu}\Phi + \sqrt{2}\zeta_{\alpha}F \qquad (329)$$

and

$$\delta F = -i\sqrt{2}\zeta_{\dot{\alpha}}\left(\overline{\sigma}^{\mu}\right)^{\alpha\dot{\beta}}\partial_{\mu}\Psi_{\beta} \qquad (330)$$

This set of equations allows us to develop a SUSY algebra characterized by the commutative properties

$$\left\{Q_{\alpha}, \overline{Q}_{\dot{\beta}}\right\} = 2\sigma^{\mu}_{\alpha\dot{\beta}}P_{\mu} \qquad (331)$$

$$[Q_{\alpha}, P^{\mu}] = 0 \qquad (332)$$

$$[M^{\mu\nu}, Q_{\alpha}] = -i(\sigma^{\mu\nu})_{\alpha}{}^{\beta}Q_{\beta} \qquad (333)$$

In equation (331), two successive SUSY-translated transformations in spacetime are signed. The operator Q_{α} is a fermionic operator generating the SUSY transformation. In SUSY theory, the energy of any non-vacuum state is a positive definite, with

$$4\left\langle\psi\middle|P^{0}\middle|\psi\right\rangle = \left\langle\psi\middle|Q_{\alpha}(Q_{\alpha})^{*} + (Q_{\alpha})^{*}Q_{\alpha}\middle|\psi\right\rangle \qquad (334)$$

In the SUSY vacuum, the vacuum energy is exactly zero

$$\left\langle 0\middle|P^{0}\middle|0\right\rangle = 0 \qquad (335)$$

which is equivalent to

$$Q_\alpha|0\rangle = 0 \qquad (336)$$

However, for phenomenological reasons, this has to be broken down at a scale that is, at least, higher than 1 *TeV*

$$\langle\rho\rangle = \frac{1}{2}\frac{1}{(2\pi)^3}\int d^3k\sqrt{k^2+m^2} = \frac{M_C^4}{16\pi^2}\left(1+\frac{m^2}{M_C^2}+\ldots\right) \approx 10^{60}\rho_{cri}, (337)$$

Here, $M_C \approx 1TeV$. We are still off by 60 orders of magnitude.

As such, the previous difficulties have to be resolved with other alternatives and from this, the question arises, can dark energy be a (scalar) field? This a fundamental question we have posed before.

We consider the total energy in the decoupling equations for ρ and p

$$\rho Q = \frac{\dot{Q}^2}{2} + V(Q), \quad pQ = \frac{\dot{Q}^2}{2} - V(Q) \qquad (338)$$

If the potential energy dominates, one can have negative pressure (as for inflation).

This results in the following considerations. Firstly, we aim to study dark energy with a time-dependent equation of state, that is to say, with a particular process of evolution described from the energy states. Since we have a microscopic model, we can consistently compute the cosmological perturbations through the backreaction of dilaton particles (gauges) in the gravitational field, the immediate effects of which can be obtained in the light-energy produced and considering the geometry of the corresponding gravitational scenario. This allows us to discuss the link with high-energy physics and to play the game of model-building. However, it is very hard to determine this with precision and it does not solve the *CC* problem. Instead of explaining $\Omega_\Lambda = 0.7$ of the critical energy density, we are back to $\Lambda = 0$, i.e. to flat space.

Likewise, starting with the action of scalar fields proposed in the previous chapter, we have

$$S=-\int d^4x\sqrt{-g}\left[\frac{1}{2}g^{\mu\nu}\partial_\mu\Phi\partial_\nu\Phi+V(\Phi)\right] \qquad (339)$$

where the energy-matter tensor

$$T_{\mu\nu}=-\frac{2}{\sqrt{-g}}\frac{\delta S}{\delta g^{\mu\nu}} \qquad (340)$$

is due to the action of (339), as follows

$$T^\mu{}_\nu=g^{\mu\lambda}\partial_\lambda\Phi\partial_\nu\Phi-\delta^\mu{}_\nu\left[\frac{1}{2}g^{\alpha\beta}\partial_\alpha\Phi\partial_\beta\Phi+V(\Phi)\right] \qquad (341)$$

Neglecting the spatial derivatives (FLRW[38]), we have

$$\rho=\frac{\dot{\Phi}^2}{2}+V(\Phi), \qquad p=\frac{\dot{\Phi}^2}{2}-V(\Phi) \qquad (342)$$

In the vacuum, the derivatives should be zero, that is to say

$$\rho=+V_{\min}(\Phi)=-p \qquad (343)$$

In the variation with respect to the field, we establish the field differential equation

$$\ddot{\Phi}+3H\dot{\Phi}+\frac{\partial V}{\partial\Phi}=0 \qquad (344)$$

which gives no spatial gradient because the FLRW scenario arises. In addition, there is a damping term because the field exists in curved spacetime.

---

[38] Friedman-Lemaître-Robertson-Walker metrics. These describe a homogeneous, isotropic, and expanding (or, otherwise, contracting) universe that is path-connected, but not necessarily simply connected.

We can postulate the presence of the scalar field Q with a runaway potential and $\Lambda = 0$. We assume that, initially, the field is subdominant, which means that considering the physics of the dark matter with w-parameters and scalar field a, we apply the third column of table 1 in chapter 1, and obtain

$$a(t) \propto t^{2/(3+3w_B)} \qquad (345)$$

which could be the equation of state for the background fluid, i.e. 1/3 or 0, as seen in table 1. Thus, the corresponding field differential equation of the form (344) will be

$$\ddot{Q} + 3\frac{2}{3(1+w_B)t}\dot{Q} - \alpha M^{4+\alpha} Q^{-\alpha-1} = 0 \qquad (346)$$

where a particular solution exists for (346), given by

$$Q(t) = C(\alpha, w_B, M) t^{[4-\alpha(3w_B+1)]/(4+2\alpha)} \qquad (347)$$

The matter due to field Q is

$$\rho_Q \propto a^{-3(1+w_Q)} \qquad (348)$$

and its corresponding w-parameter

$$w_Q = \frac{-2+\alpha w_B}{2+\alpha} \qquad (349)$$

The field tracks the background, i.e. it adapts its behaviour to the background fluid. Additionally, It scales less rapidly than the background fluid and will eventually come to dominate.

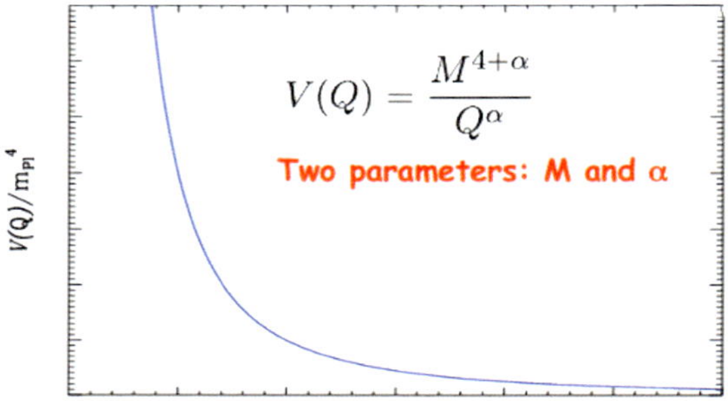

Fig. 45. Field potential per matter m_{PI} versus dark matter by field Q.

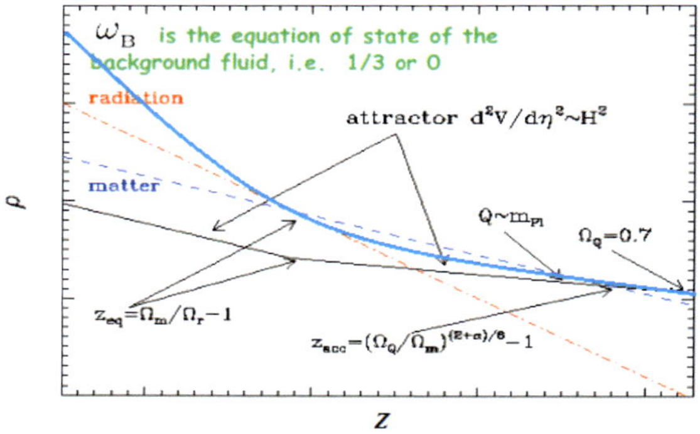

Fig. 46. *w*-parameter adjusting, considering matter, radiation, and attractor for inflation.

Of course, once the field starts dominating the matter content of the universe, the previous solution is no longer valid. This happens when

$$Z_Q = \left(\frac{\Omega_Q}{\Omega_{matter}}\right)^{(\alpha+2)/6} - 1 \cong, O(1) \qquad (350)$$

This happens precisely with the second derivative $d^2V(Q)/dQ^2$ of (346), considering the conditions of normalization to the *w*-parameter

$$\frac{d^2V(Q)}{dQ^2} = \frac{9}{2}\frac{\alpha-1}{\alpha}\left(1-\omega_Q^2\right)H^2 \qquad (351)$$

having

$$\frac{d^2V(Q)}{dQ^2} \approx \frac{\rho_Q}{Q^2} \qquad (352)$$

and

$$H^2 = \frac{\rho_Q}{m_{PI}^2} \qquad (353)$$

Today, we have

$$\langle Q \rangle_{today} \approx m_{PI} \qquad (354)$$

This particular solution is interesting because it presents an attractor that is joined for a huge range of initial conditions $10^{-37}\,GeV^4 < \rho_Q < 10^{61}\,GeV^4$.

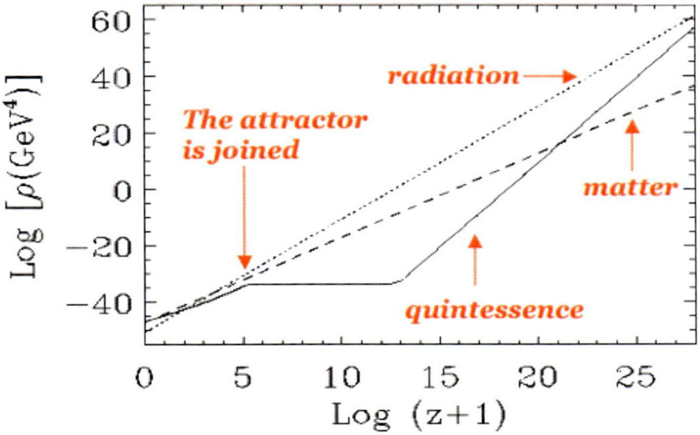

Fig. 47. $10^{-37}GeV^4 < \rho_Q < 10^{61}GeV^4$ -scenario.

This is called a "tracker field" and can also be obtained for other potential forms, for example, considering constant $Log[\rho(GeV^4)]$ after $Log(z+1) \leq 10$ and in the scale $Log[\rho(GeV^4)] = -20$ (see fig. 48).

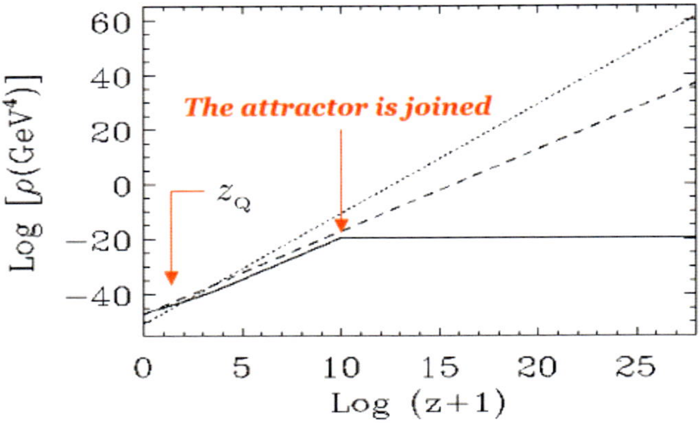

Fig. 48. With the potential from (342) under observational considerations over a short time period.

The state equation is a time-dependent (or redshift-dependent) quantity

$$w_Q = \frac{\dot{Q}^2/2 - V(Q)}{\dot{Q}^2/2 + V(Q)} = w_0 + w_1 z + \ldots (355)$$

The present value is negative and different from -1. As such, it can be distinguished from a cosmological constant (see fig. 49).

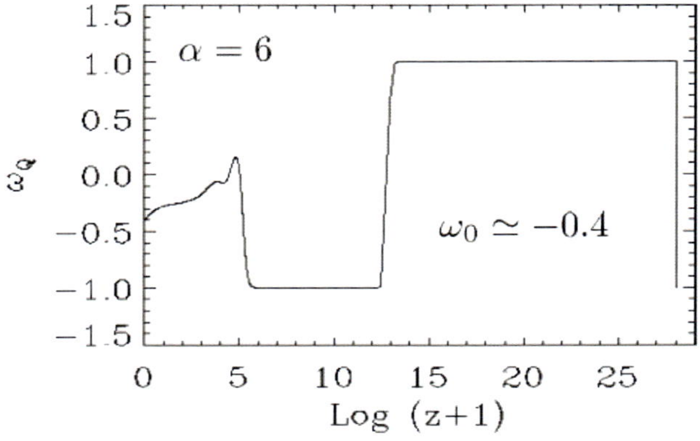

Fig. 49. Negative value of time-dependent (or redshift-dependent) quantity.

Of course, the present value of the state equation is also independent of the initial conditions (see fig. 50).

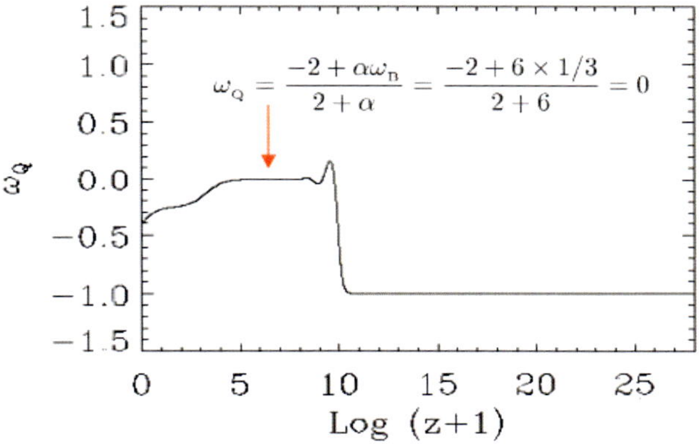

Fig. 50. Independent value from the initial conditions.

Thus, the redshift-dependent quantity with DE-perturbations and non-DE-perturbations will have for the scalar field the following phase space relationship to the attractor, as

$$w(Q) = w_0 + w_a(1-a) = w_0 + w_a \frac{z}{1+z} \tag{356}$$

Two phase-portraits are given in figures 51 and 52. Here, the WMAP was used as an astrophysically-observed fact directly measured with the background.

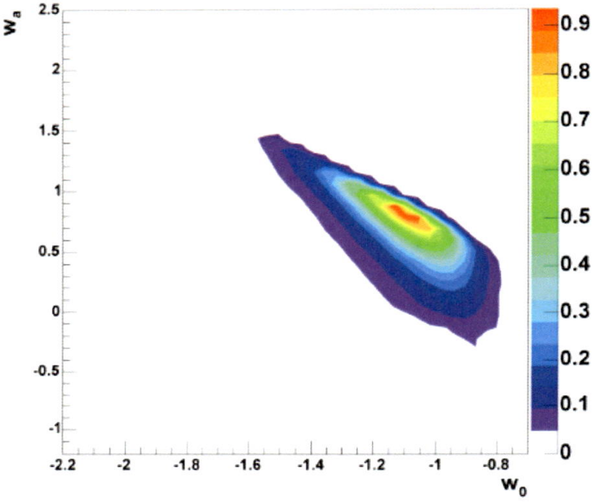

Fig. 51. Independent value from the initial conditions without DE-perturbations [75].

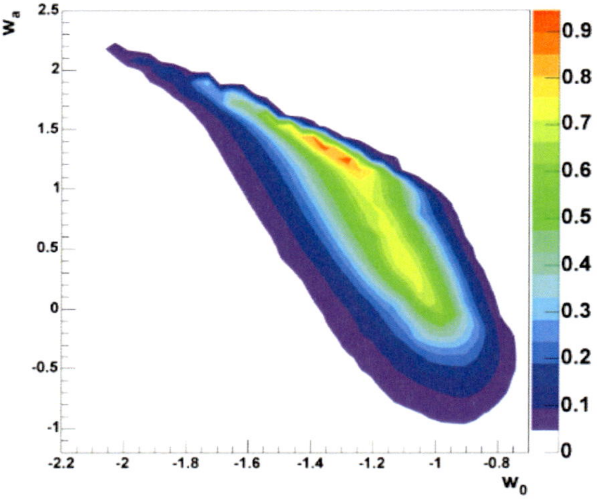

Fig. 52. Independent value from the initial conditions with DE perturbations [76].

The energy scale M of the potential is fixed by the requirement that the quintessential energy density today represents 70 % of the critical energy density, with $\Omega_{\Lambda} = 0.7$. Considering (353)-(356), we have

$$\frac{M^{4+\alpha}}{m_{PI}^{\alpha}} \approx \rho_{cri} \qquad (357)$$

then

$$Log_{10}[M(GeV)] \approx \frac{19\alpha - 47}{\alpha + 4} \qquad (358)$$

In addition, the mass of the field is tiny and of the order

$$m_Q^2 = V'' = \frac{\rho_{cri}}{m_{PI}^2} \approx \left(10^{-33} eV\right)^2 \qquad (359)$$

that is to say, a very long range force, already well constrained by various experiments (see fig. 54).

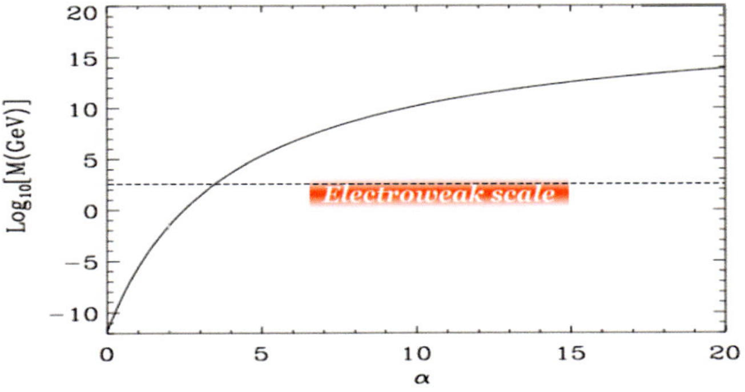

Fig. 53. Increasing matter production with respect to the electromagnetic flatness of the weak force.

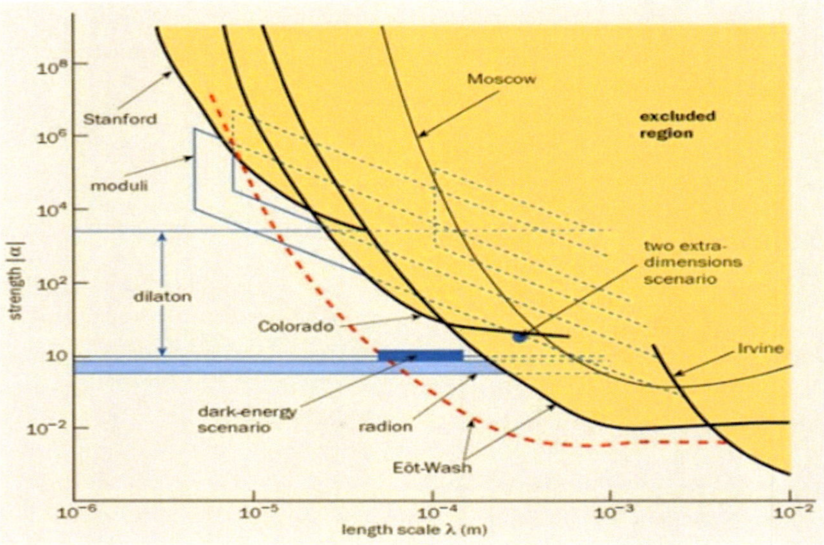

Fig. 54. Different experiments to measure the strength of α versus the length scale. In these experiments a gauge particle is used, such as a dilaton. The detected range of dark energy scenario is very short, with $\langle Q \rangle_{today} = m_{PI}$. The scale of the length appears in hte range $10^{-45} \leq \lambda \leq 10^{-33}$.

Then, the inflation indices move with the following ranges: $10^{-37} GeV^4 \leq \rho_Q \leq 10^{61} GeV^4$ and $10^{-107/\alpha} GeV^4 < \frac{Q}{m_{PI}} < 10^{-9/\alpha}$ (see fig. 55).

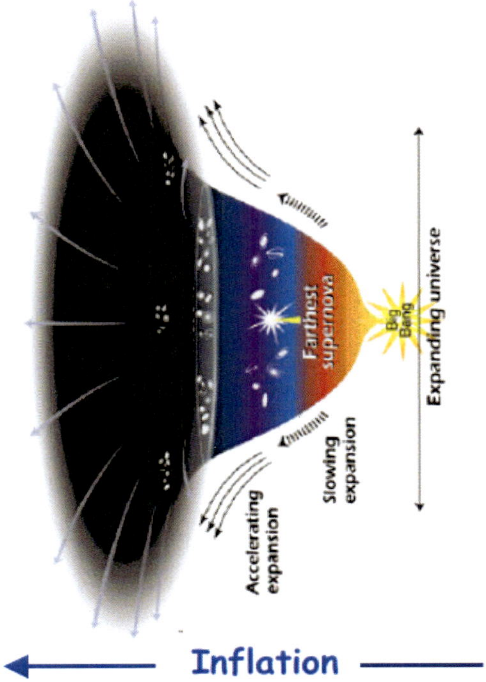

Fig. 55. Ranges of dark energy at different stages of the universe's evolution.

From this, the following questions arise:

Do we have initial conditions (final conditions from the point of view of inflation) that are compatible with the attractor? What are the initial conditions for the quintessence field at the beginning of inflation?

The answer to the first question is yes, because reheating comes afterwards, which is evidence of microscopic interactions producing inflation. To answer the second question, we need to take into account the following considerations:

i) We can test the quintessence field against the background, the evolution of which is controlled by chaotic inflation.
j) Typically, the quintessence field is frozen during inflation.

However, the stochastic effects are very important (see fig. 56) and depend on the interaction of an inflaton with the quintessence field; that is to say, from the point of view of high energy, the inflaton and the quintessence field must interact as a necessary condition for inflation to happen according to i) and j).

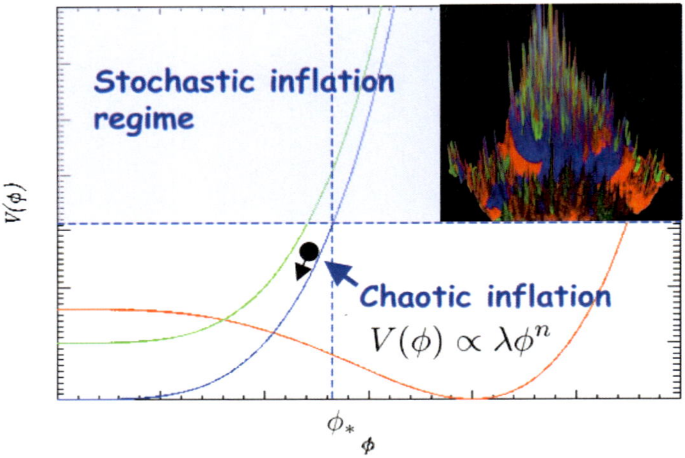

Fig. 56. Considerations on the quintessence field and its interaction with the background, the evolution of which is controlled by chaotic inflation. Stochastic effects are necessary.

The field undergoes a quantum jump $H/2\pi$ in each Hubble time. In addition, we have

$$\Delta\phi_{qu} \approx \frac{H}{2\pi}, \qquad \Delta\phi_{cl} \approx -\frac{V_{,\phi}}{3H}\Delta t = -\frac{V_{,\phi}}{3H}\frac{1}{H} \qquad (360)$$

where we can observe that

$$\Delta\phi_{qu} > \Delta\phi_{cl} \qquad (361)$$

and

$$\phi > \phi_* \approx \lambda^{-\frac{1}{n+1}} \qquad (362)$$

As such, inflation depends strongly on the quantum effects, which are important for quintessence [77]. We take $V(Q)=M^{4+\alpha}Q^{-\alpha}$ as the potential (see fig. 57) and if we consider the inflaton as a function of the quantum jump

$$\Delta Q_{qu}\approx\frac{H(\phi)}{2\pi},\qquad \Delta Q_{cl}\approx-\frac{V_{,Q}}{3H(\phi)}\Delta t=-\frac{V_{,Q}}{3H(\phi)}\frac{1}{H(\phi)} \tag{363}$$

then

$$Q>Q_{*}\approx\left(\frac{M^{4+\alpha}}{H^{3}}\right)^{\frac{1}{\alpha+1}} \tag{364}$$

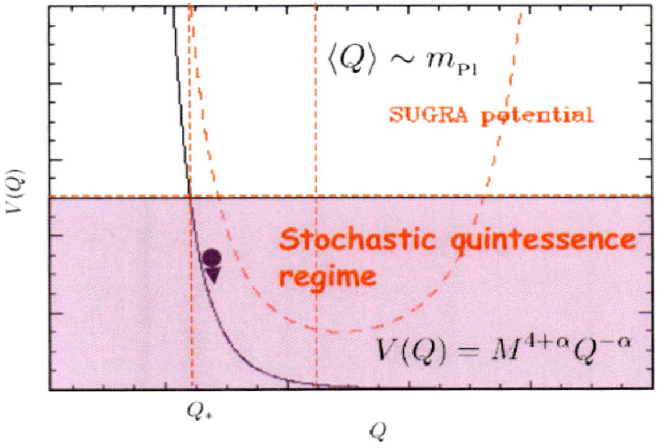

Fig. 57. Quantum effects are important for the quintessence field, which has a substantive role in inflation.

Other observations can be given in relation to figures 53 and 58 where the confidence range enlarges by the power index α and a "small" number of total e-foldings is favored. Otherwise, the quintessence field will have too much time and drift away from the allowed range of initial conditions [78].

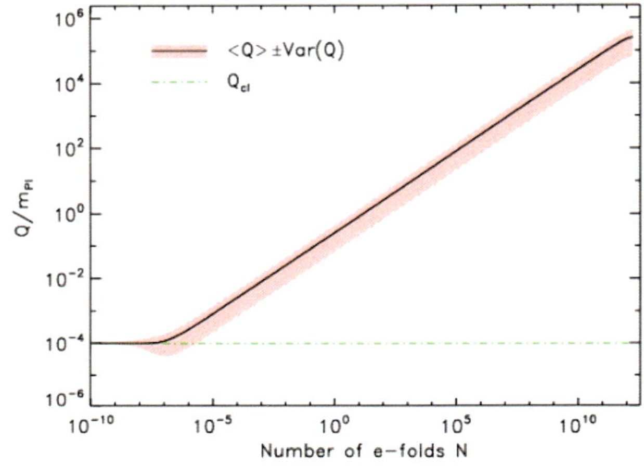

Fig. 58. *e*-folds in the moduli stack.

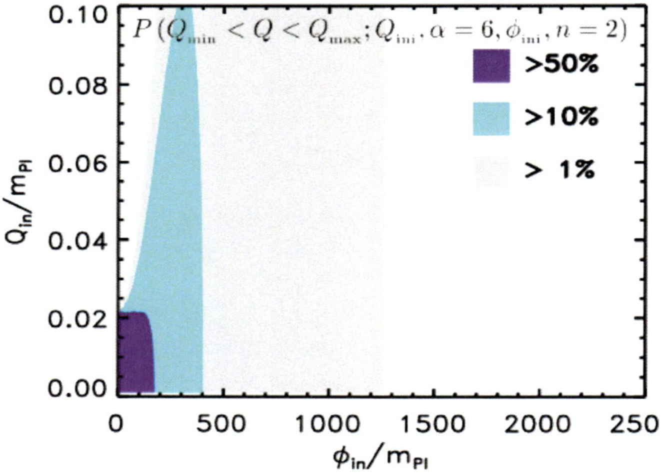

Fig. 59. $\frac{Q_{in}}{m_{PI}}$ with probability distributions of different ranges of the scalar field Q with $\alpha = 6$ and $n = 2$.

Can we find a candidate particle to be the quintessence field? In other words, can we find a candidate for the quintessence scalar field in particle physics?

Currently, there is no candidate in the Standard Model of particle physics. As such, we have to consider extensions of this model. The most popular extensions are based on supersymmetry and/or supergravity.

The following questions also arise:

Can we derive the Ratra-Peebles potential in a consistent way? What is the influence of the quantum corrections on the shape of this potential?

If dark energy is just a field, it must interact with the rest of the world. Can we compute this interaction? Can we go even further and establish a link between dark energy and string theory? Can we find a candidate with a "stringy" interpretation?

If we consider supergravity as a gauged version of supersymmetry, that is to say, when the infinitesimal SUSY transformations are made local, then

$$\delta\Phi = \sqrt{2}\xi^{\alpha}(x)\Psi_{\alpha} \qquad (365)$$

One finds in the spectrum a graviton field (*spin* = 2) and its supersymmetric partner with a field of spin 3/2, the gravitino[39]. However, can this be revealed to start the potentials of scalar fields?

We can consider the corresponding supergravity Lagrangian

$$\mathcal{L}_{\text{SUGRA}} = G_{A\bar{B}}\partial_{\mu}\Phi^{A}\partial^{\mu}(\Phi^{B})^{\dagger} + \frac{1}{\xi^{2}}e^{G}\left(G^{\bar{A}B}G_{\bar{A}}G_{B} - 3\right) + V_{D} + \ldots (366)$$

where

[39] In supergravity theories, combining general relativity and supersymmetry, the gravitino is the gauge fermion supersymmetric partner of the hypothesized graviton. It has been suggested as a candidate for dark matter.

$$G_{A\bar{B}} = \frac{\partial^2 G}{\partial \Phi^A \partial (\Phi^B)^\dagger}$$
$$\frac{\partial^2}{\partial \Phi^A \partial (\Phi^B)^\dagger}[\xi K + \ln(\xi^3 |W|^2)]$$
$$\frac{\partial^2 (\xi K)}{\partial \Phi^A \partial (\Phi^B)^\dagger} \qquad (367)$$

which represents the metric in field-space. From (366), we can identify the terms $\frac{1}{\xi^2} e^G \left(G^{\bar{A}B} G_{\bar{A}} G_B - 3\right)$ and V_D, which are known as the F-term and D-term, respectively. In the expression $\xi K + \ln(\xi^3 |W|^2)$ we identify K as the Kähler potential and W as the super-potential. The field space is a field manifold isomorphic to a sphere. Here, the SUGRA potential is no longer a positive definite!

Fig. 58. Spin-manifold.

The link between the SUGRA and the global SUSY frameworks can be sketched by

$$SUGRA = SUSY + O\left(\frac{\Phi}{m_{Pl}}\right) \qquad (368)$$

Here, we take the potential behavior given by $V(Q) = M^{4+\alpha} Q^{-\alpha}$ for the state of the scalar field $\langle Q \rangle_{today} \approx m_{Pl}$, which implies SUGRA [79]. The issue of quintessence model building must be addressed in the SUGRA framework considering the Kähler potential and super-potential. We construct the supergravity version of the Ratra-Peebles potential. A relatively general Taylor expansion leads to the Kähler potential and super-potential (for quintessence scalar field), as with the expressions

$$K_{quint} \approx XX^{\dagger} + QQ^{\dagger} + YY^{\dagger}\left\{\in_1 + \frac{(QQ^{\dagger})^q}{m_C^{2q}}\right\} + \in_2 \frac{(YY^{-})}{m_C^{2p-2}} \mathcal{K}\left(\frac{QQ^{\dagger}}{m_C^2}, \frac{XX^{\dagger}}{m_C^2}\right) + \in_3 \frac{(XX^{\dagger})^M (QQ^{\dagger})^N}{m_C^{2m+2n-2}} + \ldots \quad (369)$$

and

$$W_{quint} = \lambda X^2 Y \quad (370)$$

where m_C^2 represents the mass scale considered, i.e. the cut-off of the effective theory used. (370) can be justified if the charges of *X*, *Y*, and *Q* under *U*(1) are 1, -2, and 0. If *Y* = 0, then $\langle W_{quint} \rangle = 0$ and $\langle X \rangle = \xi$ and we have the coupling constant. Furthermore, it is necessary that $\in_1 = 0$, meaning that there is no quadratic term in *Y* to $p > 1$. In addition, $\in_3 = 0$ means that there is no direct coupling between *X* and *Q*, otherwise the matrix is not diagonal.

Realizing some straightfoward calculations, we can obtain the following expression for the potential

$$V_{quint}(Q) = e^{\xi Q^2} \frac{M^{4+2q}}{Q^{2q}} \quad (371)$$

in which we consider

$$M^{4+2q} = e^{\xi \eta^2} m_C^{2q} \lambda^2 \eta^2 \quad (372)$$

Thus

$$\eta \approx \frac{1}{\sqrt{\lambda}}\left(\frac{m_{PI}}{m_C}\right)^{q/2} \rho_{cri}^{1/4} \qquad (373)$$

and with SUGRA correction, it becomes

$$\xi Q^2 \approx \frac{Q^2}{m_{PI}^2} \qquad (374)$$

We cannot exactly recover the Ratra-Peebles potential, but we can determine the mass scale of the potential in terms of the microscopic parameters given in (372). With this simple model, we can see that the scenario still has some important problems. For example, if we have $\lambda \approx 1$, $m_C = 10^{15} GeV < m_{PI}$, and $q \approx 7$, then $\xi > 10^2 GeV$; however, to control terms like $\sum_n^{\infty} \frac{(QQ^{\dagger})^n}{m_C^{2n}}$ with $\langle Q \rangle \approx m_{PI}$, we need $\lambda \approx 1$ and $m_C = m_{PI}$, but, in some sense, the fine-tuning problem reappears!

As such, it is necessary to realize a study of the detailed corrections of SUGRA effects. For example, we can observe that the attractor solution still exists since, for large redshifts, the vev[40] of Q is small in comparison to the Planck mass. Furthermore, the exponential corrections push the state equation towards -1 at small redshifts (that is to say, as a sum of these small redshifts).

$$w_Q = \frac{\dot{Q}^2/2 - V(Q)}{\dot{Q}^2/2 + V(Q)} = w_0 + w_1 z + \ldots (373)$$

and the value of the state equation becomes "universal", i.e. it does not depend on α.

[40] Expected vacuum values.

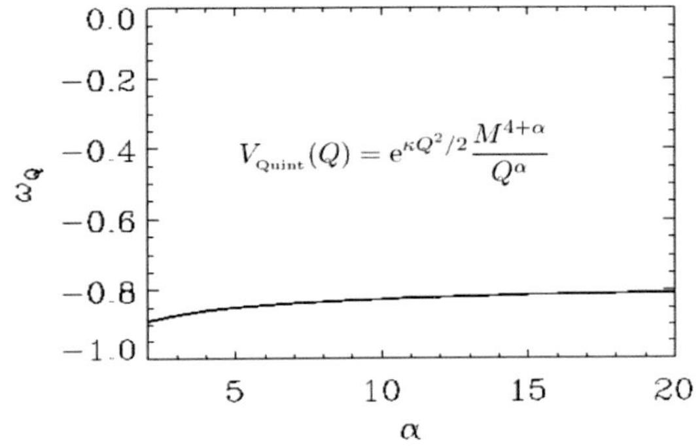

Fig. 59. The new curve describing the behavior of the quintessence scalar field depending on α.

So far, we have treated the quintessence field as being isolated from the rest of the world. The problem is not technically easy because we must direct it in the

SUGRA framework. Thus, an observable sector of mSUGRA, can be given by

$$K_{obs} = \sum_a \phi_a \phi_a^\dagger + \ldots \qquad (374)$$

and

$$W_{obs} = \frac{1}{3}\sum_{abc} \lambda_{abc}\phi_a\phi_b\phi_c + \frac{1}{2}\sum_{ab}\mu_{ab}\phi_a\phi_b \qquad (375)$$

in which the standard fields exists: electrons, quarks, dark matter, and fermions, etc. The potentials (observable potential from mSUGRA) (374) and (375) can come from the hidden sector of SUSY, with

$$K_{hid} = \sum_i z_i z_i^\dagger + \ldots \qquad (376)$$

and

$$W_{hid} = W(z_i) \qquad (377)$$

where SUSY is boken and we have Polanyi fields etc. and (376) and (377) are mediated by gravity with the risk of (374) and (375). The usual structure of the Standard Model has two sectors and taking the hidden sector of SUSY and the observable sector mSUGRA, we can modify the quintessence potential in a third sector. Here, we present the versions of the quintessence potential

$$K_{quint} = K_{quint}(d_\alpha) \qquad W_{quint} = W_{quint}(d_\alpha), \qquad \forall Q \in \{d_\alpha\} \qquad (378)$$

where the total potentials can be seen as superpositions

$$K = K_{quint} + K_{hid} + K_{obs} \qquad W = W_{quint} + W_{hid} + W_{obs}, (379)$$

Thus, the presence of dark energy will affect the electroweak transitions.

In the, mSUGRA model:

i) We have two Higgs doublets, instead of one,

ii) The EW[41] phase transition is intimately linked to the breaking of SUSY.

Without the breaking of SUSY, the Higgs potential only has a global minimum while breaking SUSY modifies the shape of the potential

$$V_{mSUGRA} = \ldots + e^{\xi K} V_{SUSY} + A_{abc}(\phi_a \phi_b \phi_c + \phi_a^\dagger \phi_b^\dagger \phi_c^\dagger) + \\ B_{ab}\left(\phi_a \phi_b + \phi_a^\dagger \phi_b^\dagger\right) + m^2_{a\bar{b}} \phi_a \phi_b^\dagger \qquad (380)$$

where $\left(\phi_a \phi_b + \phi_a^\dagger \phi_b^\dagger\right)$, which are called "soft terms" because from the point of view of particle physics, they represent soft photons as certain bosons. In the term $e^{\xi K}$ in which the potential K must be included, $W_{obs} = \mu(H_u^+ H_d^- - H_u^0 H_d^0)$ (see the geometrical scenario in fig. 60) to develop the electroweak phase transition at large Higgs masses. The

[41] Electroweak phase transition at large Higgs masses.

particles acquire mass when the Higgs masses acquire a non-vanishing vev. One gets two "types" of particle

$$m_u = \lambda_u \left\langle H_u^0 \right\rangle, \qquad m_d = \lambda_u \left\langle H_d^0 \right\rangle \tag{381}$$

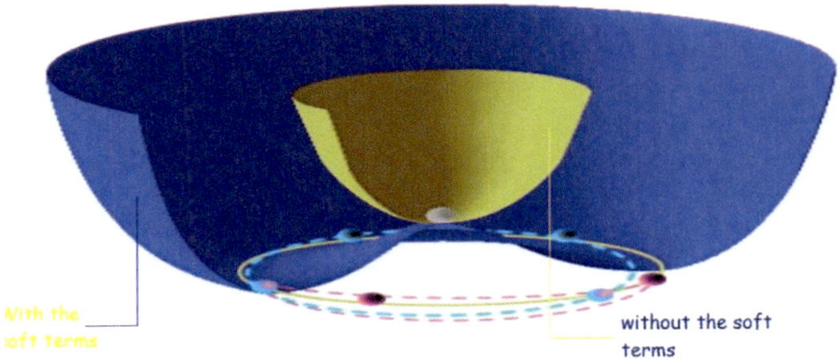

Fig. 60. Geometrical scenario to develop the electroweak phase transition at large Higgs masses.

In the presence of dark energy, the vevs of the Higgs masses become Q-dependent, that is to say, with components of vacuum expectation values

$$\upsilon_u = \frac{\upsilon \tan\beta}{\sqrt{1+\tan\beta}}, \qquad \upsilon_d = \frac{\upsilon}{\sqrt{1+\tan\beta}} \tag{382}$$

having the masses

$$m_u(Q) = \lambda_u e^{\xi K_{quint}/2 \sum_i |a_i|^2/2} \upsilon_u(Q) \tag{383}$$

and

$$m_u(Q) = \lambda_d e^{\xi K_{quint}/2 \sum_i |a_i|^2/2} \upsilon_d(Q) \tag{384}$$

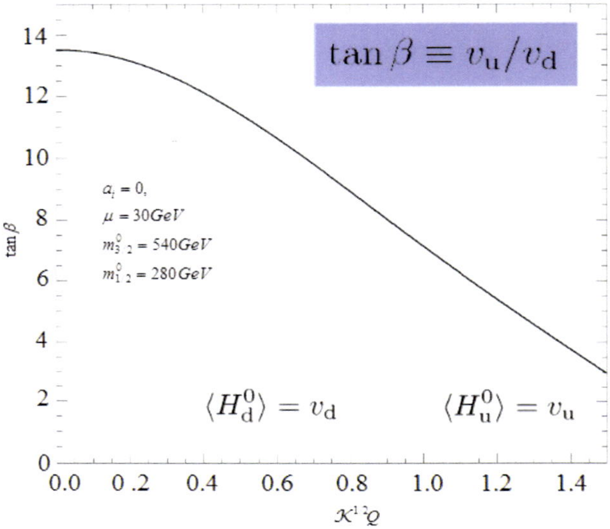

Fig. 61. The vevs of the Higgs masses derived from the SUGRA model+ *m*SUGRA.

Fermions pick up a Q-dependent mass, which is not the same for the u or d particles. This is calculable entirely from SUGRA.

Some consequences of this new formalism that evidence dark matter are:

a) The presence of a fifth force, that is to say, a parameter independent of the two vevs of the Higgs masses (see fig. 62)

$$\alpha_{u,d}(Q) = \left| \frac{1}{\xi^{1/2}} \frac{d \ln m_{u,d}(Q)}{dQ} \right| < 10^{-2.5} \qquad (385)$$

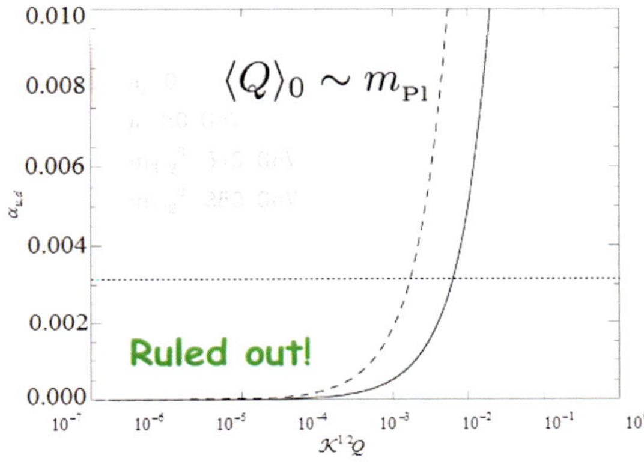

Fig. 61. Q-dependence of the two vevs of the Higgs masses.

b) Additionally, we have the violation of the (weak) equivalence principle (because there are two Higgs!). From the equation with different estimations

$$\eta_{AB} = \left(\frac{\Delta a}{a}\right)_{AB} = 2\frac{a_A - a_B}{a_A + a_B} \approx \frac{1}{2}\alpha_E(\alpha_A - \alpha_B) \qquad (386)$$

we obtain the current limits

$$\eta_{AB} = (\pm 0.1 \pm 2.7 \pm 1.7) \times 10^{-13} \qquad (387)$$

Other effects are also obtained, for example, the variation of constants (fine structure constant etc.), the proton to electron mass ratio, and the Chameleon Density model because of interactions between dark matter and energy (hence, one can have $w < -1$, meaning that the expansion of the universe is accelerating). The cosmological constant seems to be a natural candidate, but it is difficult to understand its magnitude. Even SUSY cannot really help. Quintessence is a model of dark energy where a scalar field is considered responsible for the accelerated expansion of the universe. It has some nice properties, like the ability to "solve" the coincidence problem. In addition, the quintessence equation of state is now not -1, as for the cosmological constant, and is redshift-dependent: this would be a clear and observational signature. However, implementing quintessence in high energy physics is difficult and no fully satisfactory model exists at present.

The interaction of quintessence with the rest of the world is non-trivial and can lead to interesting phenomena and/or constraints. Probing dark energy does not just involve measuring the equation of state, but all the things discussed. Local gravity tests in the solar system are also important. Consider ϕ, a "matter" field (in fact, it should be a fermion). Because of its coupling with the quintessence field, its mass can change, for example, in the action

$$S_{mat}[\phi] = -\frac{1}{2}\int d^4x\sqrt{-g}\left[g^{\mu\nu}\partial_\mu\phi\partial_\nu\phi + m^2\left(\frac{Q}{M_{PI}}\right)\phi^2\right] \qquad (388)$$

where there is variation of the mass M_{PI}. Then, through the redefinition of the corresponding space-time

$$\widetilde{g}_{\mu\nu} = A^2 g_{\mu\nu}, \qquad \widetilde{\phi} = \frac{\phi}{A}, \qquad m(Q) = \frac{A(Q)}{A(0)} \qquad (389)$$

we obtain a scalar-tensor theory (in the Einstein frame), under the action

$$S_{mat}[\widetilde{\phi}, A^2 g_{\mu\nu}] = -\frac{1}{2}\int d^4x\sqrt{-\widetilde{g}}\left[\widetilde{g}^{\mu\nu}\partial_\mu\widetilde{\phi}\partial_\nu\widetilde{\phi} + m^2(0)\widetilde{\phi}^2\right] \qquad (390)$$

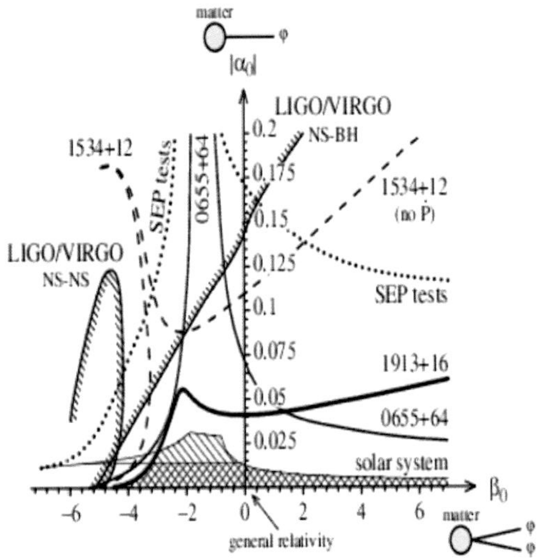

Fig. 62. Behavior of local gravity in Einstein frames for different sidereal systems with mass peculiarities.

where the term $A^2 g_{uv}$ is constrained by solar system and/or pulsar tests, having the coefficients

$$\alpha = \frac{d \ln A}{dQ}, \qquad \beta = \frac{d\alpha}{dQ} \tag{391}$$

Likewise, the evolution of small inhomogeneities is controlled by the perturbed Klein-Gordon equation

$$\delta Q_k'' + 2\frac{a'}{a}\delta Q_k' + \left[k^2 + a^2 \frac{d^2 V(Q)}{dQ^2}\right]\delta Q_k = -2a^2 \frac{dV}{dQ}\Phi_k + 4Q^2\Phi_k', \tag{392}$$

where

$$\frac{d^2 V(Q)}{dQ^2} = \frac{9}{2}\frac{\alpha+1}{\alpha}\left(1 - w_Q^2\right)H^2 \tag{393}$$

and the Jeans parameter appears, which we fully identify as the proportional inverse of the inflation

$$\ell_{Jeans} \approx H^{-1} \qquad (394)$$

We can identify this parameter in the WMAP, as in figure 63.

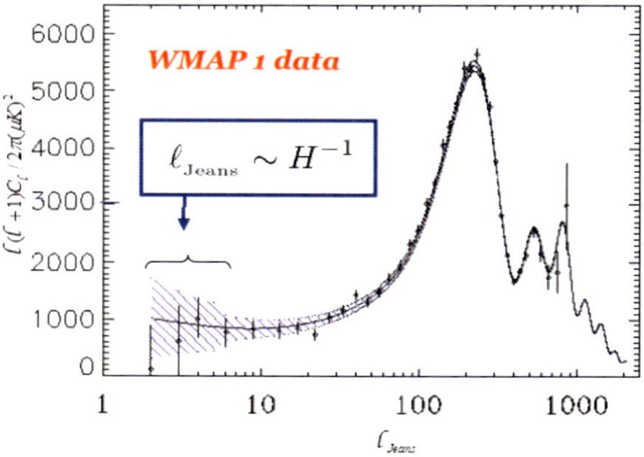

Fig. 63. Identification of the parameter ℓ_{Jeans} in a graph of temperature fluctuations with the cosmic microwave background given by the WMAP satellite.

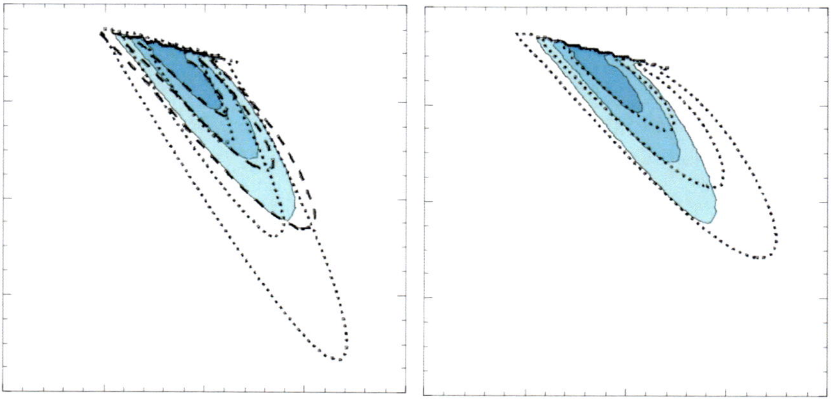

Fig. 64. Clustering of quintessence only on scales of the order of the Hubble radius.

CHAPTER 6

TECHNOLOGICAL APPLICATIONS

6.1. Field Torsion and Electromagnetic Propulsion

6.1.1. Introduction

Diverse works on electrodynamics, field observables, and other related phenomena for producing electrodynamical processes from purely electromagnetic sources (*reactor*), describe the determination of a line of electromagnetic plasma and make the hypothetical contention that a Lorentz force can be produced for the movement and propulsion of a hypothetical vehicle. Likewise, considered as a geometrical invariant and field observable, torsion can be used to design an electronic tube with an infinite cycle of electromagnetic plasma as the fundamental basis of a reactor. The curvature energy is in the form of conduction, directing and obtaining fermion species for machines that use levitation and electromagnetic impulses as part of an advanced aerospace ship.

6.1.2. Electromagnetic Plasma and Fermion Species

Electrons themselves are fermions in the material context. Likewise, the photonic quality of fermions, with photons transmitting information between electrons, determines many interesting processes from electromagnetic fields involved in an electromagnetic flow called electromagnetic plasma. In fact, electromagnetic plasma should not necessarily be considered an electromagnetic object resulting in superconduction. It can be the result of a multi-irradiative electromagnetic field, which is directed and conducted through magnets and dielectric materials, generating derived field products at the microscopic level.

These microscopic electromagnetic products are fermions, which, through their Majorana states, can produce interesting effects like electro-anti-gravity. However, this is not the principal goal of this research. Here, we are interested in the possibilities and justifications of designing and developing a reactor of an electromagnetic nature using field torsion.

In superconducting materials, Majorana fermions can emerge as (non-fundamental) quasi-particles. This is possible because a quasi-particle in a superconductor is its own anti-particle. Fermion production can be established through superconductivity with very low temperature conditions. Fermions can be produced by the microscopic theory of superconductivity and they can be separated through the application of an electromagnetic field achieving two regions with different electrical and magnetic properties.

The Shockley states at the end points of superconducting wires or line defects offer alternatives to a purely electrical source[42]. An altogether different source uses the fractional quantum Hall effect as a substitute for the superconductor. This gives us the central idea of our research: using an electromagnetic plasma line to generate corresponding polarons, which are differentiated fermion spaces that can be established in combining electro-anti-gravity effects like the levitation, suspension, and propulsion of a ship, for example. The corresponding rotation of the line (as a circular ring of plasma inside a torus) must have high velocity to generate a Lorentz force and this must be sufficiently strong to generate an impulse and move an object, likewise creating the self-generation of energy in other diverse processes such as levitation and flying. The field torsion can be used to establish and indicate the action of the Lorentz force agitating space and provoking undulation, resulting in energy that can be used for the flying and displacement of a ship.

[42] These Shockley states can determine an equivalent state to the superconducting state with electromagnetic plasma using a quantum Hall effect.

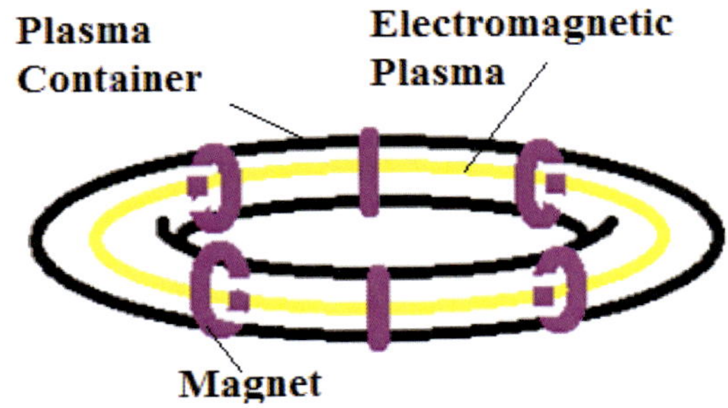
Plasma
Container
Electromagnetic
Plasma
Magnet

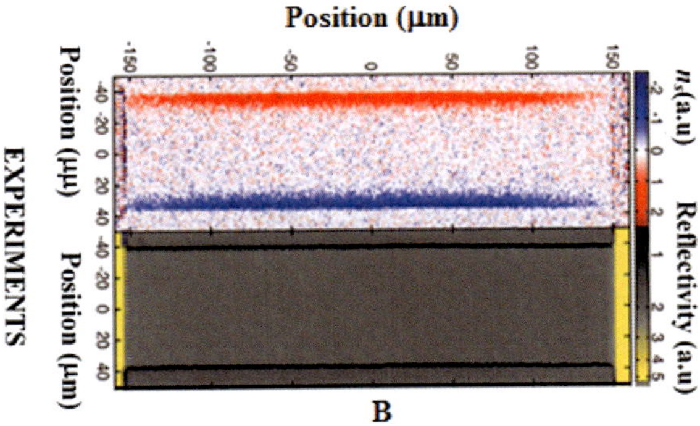
Position (μm)
Position (μμ)
Position (μm)
EXPERIMENTS
n_s(a.u)
Reflectivity (a.u)

B

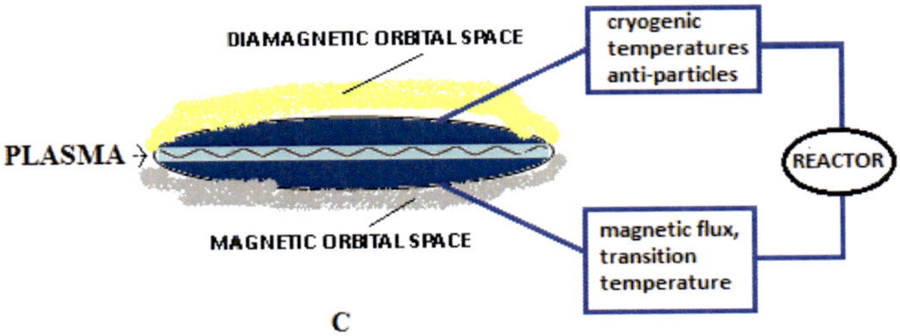
DIAMAGNETIC ORBITAL SPACE
PLASMA
MAGNETIC ORBITAL SPACE
cryogenic
temperatures
anti-particles
REACTOR
magnetic flux,
transition
temperature

C

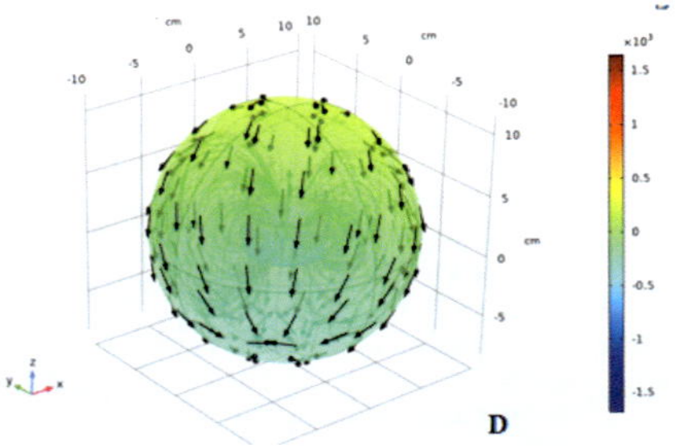

Fig. 65. A) The double magnetic layer principle stays defined with the magnetic field of each magnet and the magnetic field generated by the current in the electromagnetic plasma. B) Two differentiated fermionic (bands) orbital spaces showing the trench effect in superconduction. C) Two differentiated fermionic (bands) orbital spaces: condensation process. D) Magnetic flow with a critical temperature that is greater or smaller than the transition temperature. The impulse is obtained by the magnetic field that passes across the cylindrical ring.

6.1.2.1. Electromagnetic Plasma: Infinite Circuit

The electromagnetic plasma can be defined as the magnetic fluid $\approx \boldsymbol{II}_H(\sigma, E)$[43] (see the lemma published in [80] in advanced studies in electrodynamics). The electron flow is continuous and saturation with electrons (high electron density) occurs due to the permanent electric current. We consider a tube whose co-cycle of current has constant circulation to be an infinite circuit

$$\oiint_S \Omega_H dS = \oint_{\mathfrak{T}} \mathbf{v} dl = 0 \qquad (395)$$

The tube[44] is a device defined by a topological vector space

$$\mathfrak{T} = \{\Omega_H \in \approx \boldsymbol{II}_H(\sigma, E) | \operatorname{div}\Omega_H = 0\} \qquad (396)$$

[43] The symbol $\approx \boldsymbol{II}_H$ refers to the magnetic fluid as studied in (MHD).
[44] Here, the tube is a cylinder of infinitesimal radius (at least of a conductor defined by the plasma) joined by their extremes and forming a circular torus.

where Ω_H is the electromagnetic vorticity given by the equation

$$\Omega_H = \text{rot}\,\mathbf{v} \qquad (397)$$

where **v** is the velocity obtained by the plasma due the magnetic flow density that acts through the magnetic field lines

$$\frac{\partial \mathbf{B}}{\partial t} = \text{rot}(\mathbf{v} \times \mathbf{B}) \qquad (398)$$

Likewise, we define the reactor as the space

$$\mathcal{R} = \{ j \in \mathfrak{J} \,|\, \text{rot} H = j \} \qquad (399)$$

with a permanent current required inside it. This permanent current is obtained in the device by the double magnetic layer principle (see fig. 65 A)).

The magnetic energy acting on the plasma of current j satisfies the hypothetical magnetic state potential φ_m given for

$$\text{div}\,\text{grad}\varphi_m = \text{div}\,H = 0 \qquad (400)$$

That is to say, the magnetic intensity lines closely follow any point exterior to the electromagnetic plasma current inside the space. Now, it is necessary to establish that the torsion is implicit in electromagnetic plasma and determined by a magnetic field. This relies on a conjecture about the physical nature of torsion in the universe, which, as a second curvature, can be established as below.

Conjecture 6.1.2.1. The nature of torsion is magnetic[45].

---

[45] This may be because the magnetic interaction is born as an additional property of materials in the universe. Considering a lemma (conjecture) published in [80] that says "all fluid in the universe is magnetic", we locate the torsion as a matter-spin interaction and that the electromagnetic field exists across the whole of the universe and, as an additional ingredient for the movement to agitate space and produce

In addition, we need the following lemma.

Lemma 6.1.2.1. [MHD] The vorticity $\Omega_H = \text{rot}\,\mathbf{v}$ envelops the force lines of the induced field for media movement $\approx \boldsymbol{II}_e$ (media currents) in field *H*.

Proof. We use the Kelvin circulation theorem and Helmholtz theorem with its corresponding corollary to describe the twistor lines in the vorticity flow ◆. As such, it is necessary to enunciate the following theorem.

Theorem (F. Bulnes) 6.1.2.1. The vorticity $\Omega_H = \text{rot}\,\mathbf{v}$ involves a torsion in the space $SO(2,\mathbb{C})C(p)$ whose nature is magnetic. The circles are formed as "freeze fields".

Proof [81] ◆

Likewise, we consider annular magnetic fields defined by the magnetic contender given by the Maxwell equation $\text{div}\, H = 0$, which is established by the current inside $\mathcal{R}$ with current circles of length $2\pi a, a > 0$. Then, the magnetic field is the gradient of the magnetic potential.

$$H = -\text{grad}\,\varphi_m = -\text{grad}\left(\frac{I}{4\pi\gamma}\Sigma\right) = -\frac{I}{4\pi\gamma}\frac{d\Sigma}{dx} =$$
$$-\frac{I}{4\pi\gamma}2\pi\left[\frac{a^2}{\left(\sqrt{a^2+x^2}\right)^3}\right] = \frac{I}{2\gamma}\frac{a^2}{\left(\sqrt{a^2+x^2}\right)^3}$$

(401)

The magnetic field involves torsion and is given as

---

energy, we have the torsion. As a result, given the conjecture is true and the torsion born of the magnetic field at the microscopic level up to the intersidereal magnetic fields that give energy to a galaxy.

$$(x^2, a) \mapsto \frac{a^2}{a^2 + z\bar{z} + r^2(a + z\bar{z})^2} \qquad (402)$$

considering a trajectory with a vorticity equal to a conic spiral in decline.

6.1.2.2. Electromagnetic Plasma: Fermion Products

Purely electromagnetic plasma satisfies the homology

$$H^1(\Pi(X), \Omega) \cong H^1(P\mathcal{T}, O(-2,-2))_{46} \qquad (403)$$

with the current superconducting effect obtained in magnetic levitation. This superconducting current can be constructed though classical electromagnetism using the homology relations explained in [14, 82][47].

Likewise, considering the Lie bracket $[e_1, e_2]$, which includes the elements $e_1, e_2 \in \mathfrak{E}$ (with $\mathfrak{E}$ as a £-module and *e* as an electron [1, 2]) and as a consequence of its realization in Majorana states satifying [2, 3], we have

$$(\psi_1, \psi_2)_{\otimes} = \psi_1 \otimes \psi_2^* - \psi_2 \otimes \psi_1^* - [\psi_1, \psi_2] \qquad (404)_{48}$$

This establishes the possibility of creating a space-time region with electro-anti-gravity acting on the particles of an object and its immediate ambient surroundings induced by an electromagnetic source (reactor) wrapping the object (fermion condensation), with two clearly differentiated orbital spaces, as established in the numerical simulation in figure 65 B). The two

---

[46] Here, $\Pi(X)$ is a set of singularities that, in the prototype, concerns the source or reactor from which all the electromagnetic fields and energy of the vehicle are born. The space $O(-2,-2)$ is the homogeneous vector bundle of lines considering the lines of the magnetic field in the twistor generated by movement in the magnetic field (rotation ring of the vehicle). $P\mathcal{T}$ is a twistor generated as a vortex of the magnetic field repulsion. Ω is the corresponding differential space to integration.

[47] A Lagrangian treatment realized to start the two actions: electromagnetic action + isodynamic rotation action.

[48] These are elements of a bi-sided ideal *J* in superconduction.

orbital spaces are also included along with the application of an electric field to generate fermion spaces.

In figure 65 B), we have one orbital space $|\mathbb{P}^+|$ as a fermionic Fock space [4] without polarity, which is therefore "diamagnetic", and the other orbital space, in which the polarity is concentrated. Therefore, a certain superconductivity is found in the twistor (*torsion*), creating intense magnetism (see fig. 65 B)).

In this case, the link integrals are the same as the path integrals followed in the electron-phonon-electron diagram (see fig. 66 A), B)), as studied in BCS-theory, and their integrant can be an inner product (*Hermitian*) of the two electromagnetic signals derived from the super-electrons e^*.

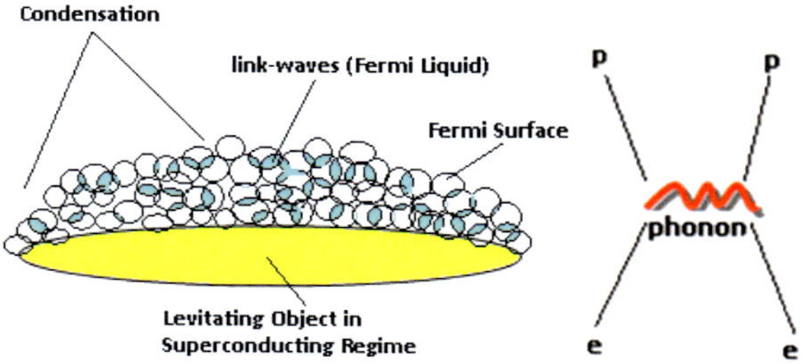

A B

Fig. 66. A) Fermi condensation resulting in electromagnetic effects acting on an object in the superconducting regime. B) Link-wave of electron-phonon-electron.

If Ω is a domain of the space-time affected by the "polarons"[49] created by the interaction between electron-phonon (figs. 4 A), B)), their Majorana states satisfy the integral given by [6]. Then, to obtain an electro-anti-gravity structure for the object, it is necessary to apply an algebraic structure in QED [1, 2] that satisfies a certain Lagrangian density $\mathcal{L}$ of the

[49] A *polaron* is a quasi-particle used in condensed matter physics for understanding the interactions between electrons and atoms in solid material.

superconductor fields from the products and the products of their Majorana states $\psi \otimes \psi^*$ [2-4, 6, 8].

As mentioned previously, the principal goal is to obtain purely electromagnetic modes and waves through electromagnetic plasma produced in a tube equalizing the modes obtained by superconductivity. The geometry of the reactor design is not arbitrary, but obeys the research given in [jeema1, jemma2].

Remember that the super-current in the presence of a magnetic field of the vector potential $\vec{A}$ is given by

$$J_S = \frac{e^*\hbar}{2m^*i}(\xi^*\nabla\xi - \xi\nabla\xi^*) - \frac{(e^*)^2}{m^*c}|\xi|^2\vec{A} \qquad (405)$$

with some works considering each electron as a Faraday form, with Stokes integration relating the current to the magnetic flow in a closed plasma circuit, the universal tensor product property, Ginzburg-Landau theory, and the Hermitian form can demonstrate that the current (405) from the electromagnetic ideal $\mathfrak{E}\otimes\mathfrak{H}$ (established in table 1, page 258 [14]) whose realizations are the states in (10).

A magnetic flow with a critical temperature of less than the transition temperature, and the torsion of the magnetic field, we can obtain an impulse mechanism (see fig. D).

6.1.2.3. Curvature Energy for Controlling Electromagnetic Plasma

We consider the plasma linearized movement equation [8]

$$\rho_m \frac{\partial \mathbf{v}}{\partial t} = \boldsymbol{j}\times\boldsymbol{B} - \nabla P - \rho_m\nabla\phi \qquad (406)$$

Then, we have the following result considering lemma 2.1 with identity (1) for the plasma physics of electromagnetic plasma

$$\approx \boldsymbol{II}_e(n\mathbf{v}, \boldsymbol{j}) = \approx \boldsymbol{II}(\boldsymbol{j}, \rho_m)$$

Theorem (F. Bulnes) 6.1.2.3.1. Let $\approx \boldsymbol{II}_H(\sigma, E)$ be the plasma. We consider equation (17) with forces on electrical charges q and mass m in the Gauss unit system. Its curvature is given by

$$\kappa = \frac{q}{mc}\left|\boldsymbol{H}\right| \qquad (407)$$

***Proof* [81]** ◆

Likewise, for a convenient torus coordinate system, we can deduce the proper magnetic field ***H***, which gives form to the electromagnetic plasma in an infinite circuit[50]. In addition, considering [12, 13] (see fig. 5), we can characterize the magnetic force in the torus coordinates ζ as $F(\zeta) = H(\zeta)p(\zeta,\eta)$ where $H(\zeta)$ is the integral along a field line derived from the scalar pressure equilibrium $\nabla P = \frac{1}{c}\boldsymbol{j} \times \boldsymbol{B}$ and $p(\zeta,\eta)$ is a pulse signal having

$$A(\omega) = \int_{-\zeta*}^{\zeta*} F(\zeta)e^{-j\omega\zeta}d\zeta \qquad (408)$$

Thus, we can obtain an infinite cycle with torus geometry of electromagnetic plasma and considering the torsion and curvature established.

If, initially, the velocity has a component parallel to ***H***, this component will be constant and the movement is the composition of two movements. There is a rotation around the field direction of the field and a translation in the stated direction (see fig. 4).

---

[50] As was suggested in theorems 2.1 and 3.1 in this work.

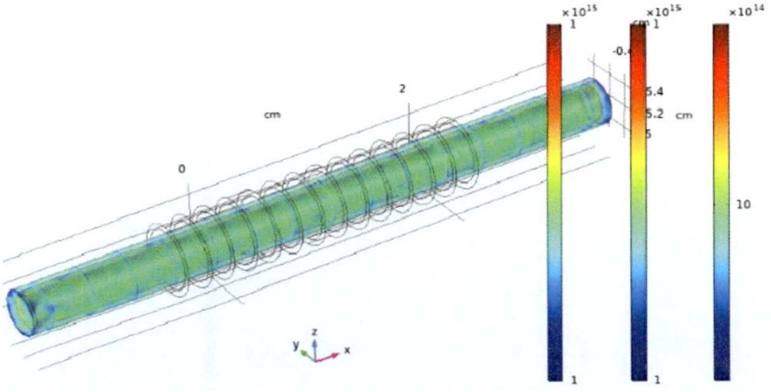

Fig. 67. 3-D [81] linearized model of electromagnetic plasma [7].

We realized some experiments of electronic homology to produce similar effects of electromagnetic plasma under certain currents and voltages.

To produce electromagnetic plasma in the laboratory for experimental purposes, we used capacitors of $100\mu f$ at 25 Volts in a parallel disposition to work the electromagnetic coil (see fig. 68 A), B)).

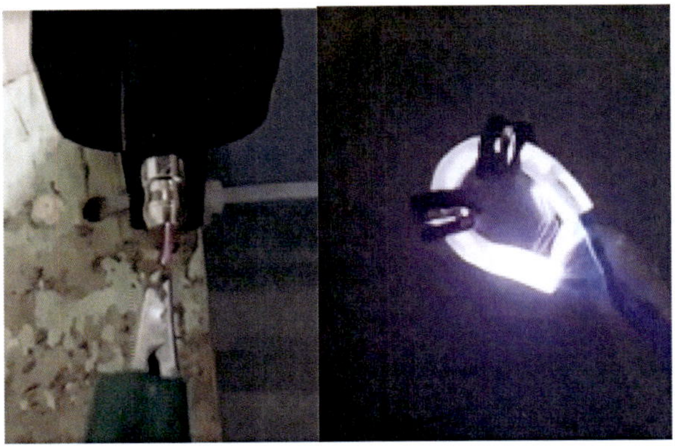

A B

Fig. 68. A) Obtaining electromagnetic plasma in an electronics laboratory. B) Electromagnetic plasma in a round container of crystalline material [81].

The reactor could be the central motor of a flying vehicle, which obtains derived products from electromagnetic plasma via the Lorentz force, and other products from several and different drifting effects and the torsion effect (see fig. 69).

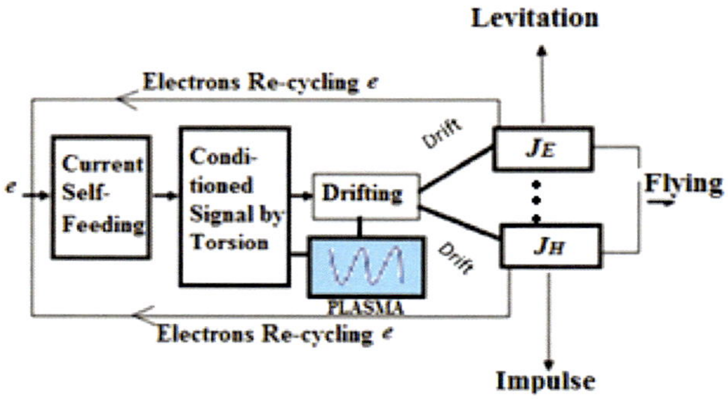

Fig. 69. Reactor block diagram [82-85].

To obtain a reactor design using electromagnetic plasma, we need to establish certain invariants and the intrinsic properties of the plasma as a magnetic fluid before we consider it as an electronic fluid. This leads us to establish certain geometrical invariants, curvature and torsion, born of field theory as observables, but that determine the geometry of the plasma under magnetic field conditions. Likewise, using the MHD we establish the construction of an infinite cycle (circular plasma torus), which is its natural geometry due to the nature of torsion in vorticity phenomena, and is observed in the different states of some sidereal objects, such as galaxies in their early phase. However, going beyond astrophysical considerations, we can characterize electromagnetic plasma as a topological space $\approx \boldsymbol{\Pi}_H(\sigma, E)$, with which we establish torsion in the space $SO(2,\mathbb{C})C(p)$, (rotations on an inverted cone) whose nature is magnetic. Considering equation (17), we can establish the expression of curvature in plasma, which suggests other effects due to torsion in the field context and under magnetic field conditions with respect to velocity, with varied and different drifts and different particle species, as mentioned. This could result in the generation of effects like diamagnetism and magnetic twistors etc., generated from equation (407) for different dispositions of the velocity field with respect to the magnetic field. Finally, the curvature energy and its torsion energy can be useful for the development of signal conditioning of cyclotron frequencies and its electronic manipulation to control plasma in a purely electromagnetic treatment of this. The experiments realized demonstrate, up to a certain point, and verify, in large measure, the results in an analytical way. A Lorentz force can also be characterized in a natural way from this perspective.

6.1.2.4. Nanomedicine and Radionics

The application of torsion fields in medicine is promising. A cure for everything can be possible using photonics devices (including devices that cure all sickness, as with my hypothesis [86, 87] given to explain the appearance of organic illness and the recomposition of fields by codes given by path integrals (Bulnes-Feynman integrals [86] to improve health)).

Could it be that the universal field is the master control of our minds? If this is true, then good management of photons as bosons of a certain class could be the solution of all mental illness.

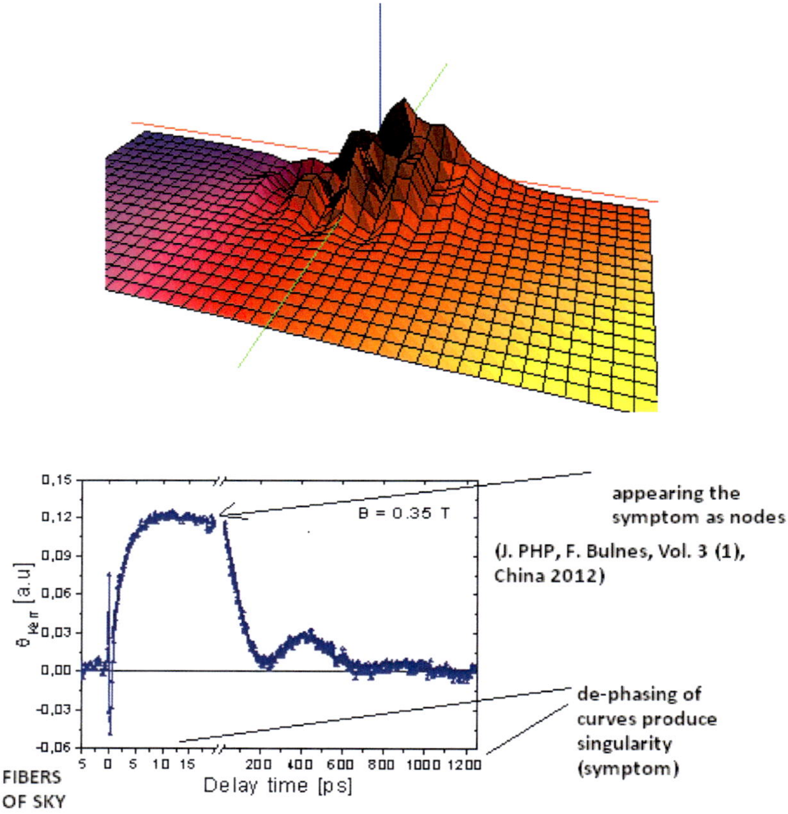

Fig. 70. What happens with the symptom space?

One of my postulates and principles for formal engineering theory is that we see the creation of technology using energy. In this, all is energy and, as I demonstrated in my work, the universe is all made up of energy because the UNISPACE or universal covering is energy. The realised argumentations were done for global analysis that use the universal covering of a Lie group with causality (or causal structure).

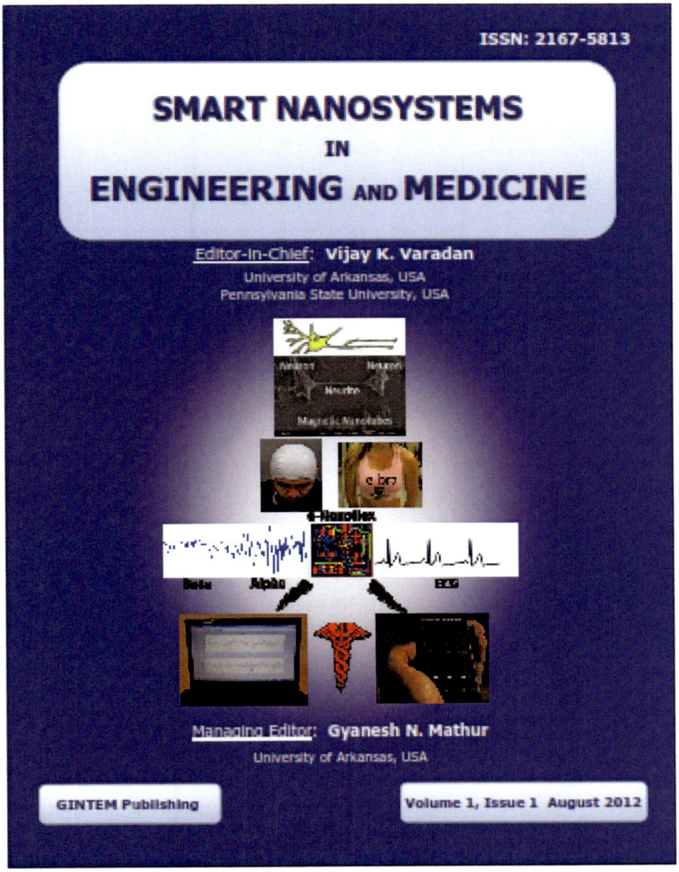

Fig. 71. Journal of Nanosystems, USA (the editor-in-chief is the prestigious neurologist Dr. Vijay Varadan, of Arkansas University [75]).

This study consider the causal structure of scattering phenomena through past and future light cones, creating the possibilities of energy for one thousand years, perpetual motion machines, and "stargates" (wormholes in space) with advanced analogues, such as synchrotronic propulsion (advanced spaceships) and disintegrative mass weapons, using the same principles in all cases.

6.1.2.5. Quantum Communication and Teleporting (Quantum Intertwining)

In the usual communication processes, the transmission of electromagnetic waves is realised with limitations due to scattering in continuum media and interference appearing for other waves and signals. In addition, we see pollution by free electrons, the molecular absorption of air, or defective routing of the TEM and TMM, as well as their secondary modes, this last case being optical fibre. Effective transmission through impermeable media, or non-permissible or invasive media, such as communication through ambient continuum media, saturated continuum media in communications, or in the presence of multi-radiative objects of large density, etc.

Likewise, the continuum transmission media produces a limitation that depends on the macroscopic characteristics of space-time such that the electromagnetic fields suffer refraction and reflection due to the substance of the proper media. To eliminate this direct action of media in signal transmission, it is necessary that our signal communication concept does not depend on the media characteristics, even if they are not "accessible or available" in the dimension of usual space-time where we have these continuum media phenomena. For example, gravitation, in the macroscopic sense, interferes in ordinary communications with the field observables of Einstein space (as could be the case of the curvature and torsion of space-time, seen as observable gravitational effects or electromagnetic scattering) acting as distortion waves and producing a field radiation due the background of the universe. Then, the signal transmission must have invariance in the electromagnetic fields, but with quantization of these fields in a space in which their transmission is topologically viable [89].

Considering the QFT (Quantum Field Theory [88]) with the second duality principle, that is to say, the wave/field duality applied to certain fermions and bosons, we can give a good fundamental explanation of the possible behaviour of these particles and their action with the intention of the communication [90]. However, it remains necessary to consider the invariance of the Maxwell equations in quantum electrodynamics (QED) [87, 91], with which we can design a form of quantum communication that establishes new paradigms of information transmission through quantum waves. In this new communication process, the intertwining concept of the particles [92], which will transmit information via the chaining of photons, using gauge bosons as supports (selectors or linkers) of the electrodynamic space at the quantum level, is required. The material part of the transmission is generated in the transceptors in their solid state. Using electrons as a solid-state source to generate photons, through a process derived from a specific spintronics devices as dots and magnetons etc., the bosons can be conformed to the information in a waveform while the plasmons constitute the quantum transmission media of these information waves. Gauge bosons will also be necessary as transmission nodes. Voice communication would be obtained by introducing phonons [92] into the photonic wave. We can establish some fundamental precisions using quantum mechanics to explain the intertwining of these particles. We must consider that for this intertwining to happen there must be more dimensions than the Minkowsky space-time dimension.

CHAPTER 7

GENERAL CONCLUSIONS

7.1. Analysis and Reflections

Understanding the role of field theory and, in particular, of quantum field theory, is important in the study of the universe as well as the varied phenomenology that occurs in relation to gravity, electromagnetism, their weak and strong interactions, and the geometrical scenario that is produced at the microscopic level in the different evolutionary steps of space-time. One of these geometric characteristics is torsion, which plays a fundamental role in the evolution and development of the universe; its objects, including galaxies, stars, supernovae, pulsars, black holes and all the phenomenology in space-time with torsion as a field observable and also the interaction between fundamental fields; and gravity and electromagnetism, ranging from the microscopic level of the action of the matter-energy tensor up to the magnetic inter-sidereal action that forms galaxies by accretion, rotation, and evolution.

All experimentally known symmetries in nature relate bosons to bosons and fermions to fermions. Theorists have hypothesised the existence of a type of symmetry, called supersymmetry, which relates bosons and fermions. Possibly, in this new interaction (mixed boson-fermion classes) the production of mass can satisfactorily explain inflation, background warming, and other phenomenological aspects of the univese.

New possibilities considering supersymmetry as a potential solution to many current problems in physics have arisen. For example, the hierarchy problem of the Standard Model—why the mass of the Higgs boson is not radiatively corrected (under renormalisation) to a very high scale such as the grand unified scale or the Planck scale—can be resolved by relating the Higgs field and its superpartner, the Higgsino. In this solution, radiative corrections due to Higgs boson loops in Feynman diagrams are cancelled out by corresponding Higgsino loops. Supersymmetry also offers answers to the grand unification of all gauge coupling constants in the Standard Model, as well as to the nature of dark matter.

Specialized studies on dark energy, and even some parallel mathematical results obtained by myself, considering the step of hadron proliferation and baryogenesis, lead me to conclude that torsion has an important role through spinor-waves as undulations in spacetime, from which gravitational waves could derive in a purely physical sense [9].

ϕ^4 -theory, QED, and QCD, as well as the whole Standard Model all assume a (3+1)-dimensional Minkowski space (3 spatial and 1 temporal dimensions) as the background against which quantum fields are defined. However, QFT a priori imposes no restriction on the number of dimensions nor the geometry of spacetime.

In condensed matter physics, QFT can be used to describe (2+1)-dimensional electron gases. In high-energy physics, string theory is a type of (1+1)-dimensional QFT, while Kaluza-Klein theory uses gravity in extra dimensions to produce gauge theories in lower dimensions.

This creation of other spacetimes encourages us to construct a new metrics to raise and lower the indices of spacetime in corresponding Lagrangians of energy balance and can help us identify or postulate the field elements responsible for various phenomena, which, until now, have not been satisfactorily explained.

However, it seems that the concept of torsion may provide the necessary link to explain all phenomena and relate all universal theories with an appropriate framework of twistor geometry, the topological quantum field theory, and other exceptional theories that formulate it as a geometrical and physical invariant.

APPENDICES

Appendix A. Photons in Quantized 6-Flat Maxwell Fields

A.1. Introduction

A review of the path integrals is realized in the quantized ambit and the scalar theory to QED, passing though the canonical quantization. Rules that well define the six-flat Maxwell fields are obtained for their propagators. Fundamental degrees of freedom are obtained from the 1-forms $A_\mu(x)$ and the strength tensor $F_{\mu\nu}$

$$F_{\mu\nu} = \partial_\mu A_\nu - \partial_\nu A_\mu = \begin{pmatrix} 0 & E^1 & E^2 & E^3 \\ -E^1 & 0 & -B^3 & B^2 \\ -E^2 & B^3 & 0 & -B^1 \\ -E^3 & -B^2 & B^1 & 0 \end{pmatrix}, \qquad (A.1)$$

which can be written in the equivalent differential form

$$F = dA, \qquad (A.2)$$

Then, the Maxwell equations remain as

$$\in^{\sigma\lambda\rho\nu} \partial_\lambda F_{\mu\nu} = 0 \qquad (dF = ddA = 0), \qquad (A.3)$$

and

$$\partial_\mu F^{\mu\nu} = j^\nu, \qquad (A.4)$$

with j^{ν} as an external source. From the action in [94]

$$S = \int d^4x \left[-\frac{1}{4} F_{\mu\nu}(x) F^{\mu\nu}(x) - j^{\nu}(x) A_{\mu}(x) \right], \qquad (A.5)$$

whose gauge transformations are determined by

$$A_{\mu} \to A'_{\mu}(x) = A_{\mu}(x) - \frac{1}{e} \partial_{\mu} \alpha(x), \qquad (A.6)$$

having $S[A'] = S[A]$, when $\partial_{\mu} j^{\mu}$.

We describe the geometry of $A_{\mu}(x)$, which is very useful for non-Abelian theories and electromagnetism defined in curved spaces. The first geometrical stack of the "bulk" of the fields $A_{\mu}(x)$ is considered the *complex line bundle*, which can be defined in a Riemannian manifold to model a spacetime in which charges can exist as sources whose fields are described for $A_{\mu}(x)$.

In this complex line bundle, which we call $\mathbb{L}$, a copy of $\mathbb{C}$ exists for each $x \in M \cong \mathbb{R}^4$ (*M* spacetime) and locally

$$\mathbb{L} = U \times \mathbb{C}, \qquad (A.7)$$

where *U* is an open region in *M*. Globally, they need not have a product structure[51].

If we consider the local coordinates (x, z) with $x \in U$ and $z \in \mathbb{C}_x$, the elements in (A.7) can be viewed as in figure A.1.

---

[51] One analogue can have this, with the Möbius strip, which is isomorphic to the bundle $\mathbb{L}$ over S^1.

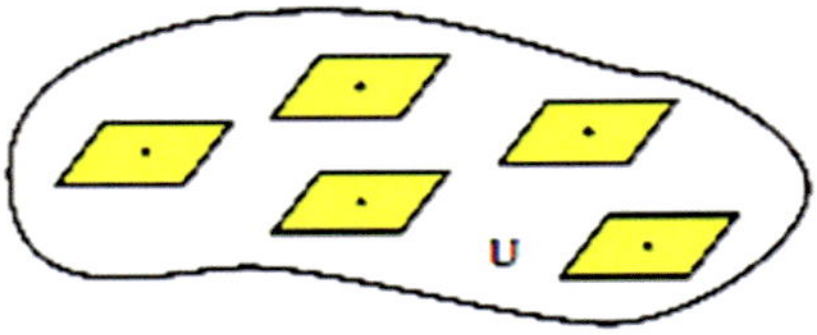

Fig. A.1. Charts of $\mathbb{L}$ over curved space *M*

A section of this line bundle gives the map

$$\sigma : M \to \mathbb{L}, \qquad (A.8)$$

$\forall \sigma(x) \in \mathbb{C}_x.$ Physically, this describes the charged particle. The local description of the section is given by the charts $(x, z(x))$. Then, the connection given by the 1-form A_μ gives the following advantages:

a) Provides derivatives of sections, that is to say

$$D_\mu z(x) = \partial_\mu z(x) + ieA_\mu(x) z(x), \qquad (A.9)$$

b) Relates neighboring fibers $\mathbb{C}_x$.

c) This connection is similar or equivalent to the Christoffel connection in general relativity.

The analysis developed through local coordinates establishes that the change in local coordinates by gauge transformations is given by

$$z'(x) = e^{i\alpha(x)} z(x), \qquad (A.10)$$

If $D'_\mu z' = e^{i\alpha} D_\mu z,$ then by (A. 9) we have

$$(\partial_\mu + ieA'_\mu)e^{i\alpha} = e^{i\alpha}(\partial_\mu + ieA_\mu), \qquad (A.11)$$

and
$$i\partial_\mu \alpha + ieA'_\mu = ieA_\mu, \qquad (A.12)$$

where

$$A'_\mu = A_\mu - \frac{1}{e}\partial_\mu \alpha, \qquad (A.13)$$

We may consider the gauge group as the set of points $\{e^{i\alpha(x)}\}$, that is to say $U(1)$, as a gauge group of G, representing electromagnetism plus other forces implying the non-Abelian nature of G.

If z has charge e, then z^* has charge $-e$. Thus, the gauge transformation takes the form of the conjugated element

$$z^{*\prime} = e^{-i\alpha(x)} z^*, \qquad D_\mu z^* = (\partial_\mu - ieA_\mu), \qquad (A.14)$$

The action S, defined by (A.5), is invariant if its net charge is 0, e.g. $(D_\mu z^*)(D^\mu z)$, which is a classically charged scalar field + electromagnetism. Thus, we know

$$S = \int \left[D_\mu \phi^* D^\mu \phi - m^2 \phi^* \phi \right] - \frac{1}{4} F_{\mu\nu} F^{\mu\nu}, \qquad (A.15)$$

where[52]

$$D_\mu \phi^* D^\mu \phi = \partial_\mu \phi^* \partial^\mu \phi - A_\mu (-ie(\partial_\mu \phi^*)\phi + ie\phi^* \partial_\mu \phi) + e^2 A_\mu A^\mu \phi^* \phi, \qquad (A.16)$$

At this point we will begin our treatment.

---

[52] Note that j^ν from (A.5) has as its source the states ϕ, charge e, and quanta $Q = \pm e$.

A.2. Path Integral Quantization of A_μ

Expressing the action (A.5) as a quadric form on A_μ, we have

$$S = -\frac{1}{4}\int d^4x(\partial_\mu A_\nu - \partial_\nu A_\mu)(\partial^\mu A^\nu - \partial^\nu A^\mu)$$
$$= \frac{1}{2}\int d^4k A_\mu(k)[-k^2 g^{\mu\nu} + k^\mu k^\nu]A_\nu(-k), \qquad (A.17)$$

with the expectation that

$$< A_\mu(k), A_\nu(k) >\approx i\times[-k^2 g^{\mu\nu} + k^{\mu\nu}]^{-1}\times\delta^4(k+k'), \qquad (A.18)$$

An interesting problem is

$$[-k^2 g^{\mu\nu} + k^\mu k^\nu]k_\nu\alpha(k) = (-k^2 k^\mu + k^2 k^\mu)\alpha = 0, \qquad (A.19)$$

This is not the inverse, since $\int \mathcal{D}A$ is divergent. Using this problem, from the gauge invariance, we can establish that $\delta S = 0,$ when $\delta A_\mu = \partial_\mu \alpha.$ Therefore, it is necessary to fix the gauge.

Indeed, restricting to A_μ, we have $G(A) = 0,$ that is to say, $G(A) = \partial_\mu A^\mu,$ which is the Lorentz gauge.

With this specific gauge, we can characterize the electromagnetic field (or *classical charged scalar field + electromagnetism*) through the *Faddeev-Popov quantization*[53] [95].

---

[53] Every gauge field has an associated ghost and where the gauge field acquires a mass via the Higgs mechanism, the associated ghost field acquires the same mass (in the Feynman-'t Hooft gauge only, this is not true for other gauges). In Feynman

We define $A_\mu^\alpha = A_\mu + \frac{1}{e}\partial_\mu \alpha.$ Formally

$$1 = \int \mathcal{D}[\alpha(x)]\delta(G(A^\mu))\det\left(\frac{\delta G(A^\alpha)}{\delta\alpha}\Big|_{G]=0}\right), \qquad (A.20)$$

Realizing certain work, we have

$$G(A) = \partial_\mu A^\mu - \omega(x), \qquad \frac{\delta G(A)}{\delta \alpha} = \frac{1}{e}\partial^2, \qquad (A.21)$$

and characterize our gauge as a $U(1)$-linear globally good gauge. Then, from (20), we have

$$\begin{aligned}
\int \mathcal{D}Ae^{iS[A]} &= c\int \mathcal{D}\alpha \int \mathcal{D}Ae^{iS[A]}\delta(G(A^\alpha)) \\
&= c\int \mathcal{D}\alpha \int \mathcal{D}A^\alpha e^{iS[A]}\delta(G(A^\alpha)) \qquad c = \det\left(\frac{1}{e}\partial^2\right), \\
&= c'\int \mathcal{D}\omega \int \mathcal{D}Ae^{-i\int d^4x\frac{\omega^2}{2\xi}} e^{iS[A]}\delta(\partial^\mu A_\mu - \omega) \\
&= c'\int \mathcal{D}Ae^{iS[A]}e^{-i\int \frac{1}{2\xi}(\partial_\mu A^\mu)^2},
\end{aligned}$$

using $S[A^\alpha] = S[A]$ and $\mathcal{D}A = \mathcal{D}A^\alpha$. Here,

$$c' = c \times \left(\int \mathcal{D}A\alpha\right) \times \left(\int \mathcal{D}\omega e^{-i\int \frac{1}{2\xi}(\partial_\mu A^\mu)^2}\right), \qquad (A.22)$$

---

diagrams, the ghosts appear as closed loops, wholly composed of 3-vertices and attached to the rest of the diagram via a gauge particle at each 3-vertex. Their contribution to the S-matrix is exactly cancelled (in the Feynman-'t Hooft gauge) by a contribution from a similar loop of gauge particles with only 3-vertex couplings or gauge attachments to the rest of the diagram. (A loop of gauge particles not wholly composed of 3-vertex couplings is not cancelled by ghosts.) The opposite sign of the contribution of the ghost and gauge loops is due to them having opposite fermionic/bosonic natures.

So, in the ξ-gauge, the electromagnetic action S_{EM} corresponds to

$$S_\xi = \frac{1}{2}\int d^4kA_\mu(k)[-k^2g^{\mu\nu} + \left(1-\frac{1}{\xi}\right)k^\mu k^\nu]A_\nu(-k), \qquad (A.23)$$

showing that

$$\int \mathcal{D}Ae^{iS_{EM}[A]} = c'\int \mathcal{D}Ae^{iS_\xi[A]}, \qquad (A.24)$$

where (A.23) is a force.

What then can we affirm to the integral $\int \mathcal{D}Af(A)e^{i(S_{EM}[A]+S_I[A])}$?

The same argument is correct if $f(A)$ and $S_I(A)$ are gauge invariants. To show this, we may need involve the forms $\phi,\ldots$ and we also need $\mathcal{D}\phi\mathcal{D}\phi^* = \mathcal{D}\phi^\alpha\mathcal{D}(\phi^\alpha)^*$, which is correct if *G* is unitary.

Realizing some inversions as the inversion of S_ξ, and considering the pertinent unitary transformations, we have

$$[U^2g^{\mu\nu} + AU^\mu U^\nu][CU^2g_{\mu\nu} + DU_\mu U_\nu] = CU^4\delta^\mu_\lambda + [AC + D + AD]U^2U^\mu U_\lambda,$$

Then, for the inverse we have

$$C = \frac{1}{U^4}, \qquad D = \frac{-A}{U^4(1-A)}, \qquad (A.25)$$

where $A = -1 + \frac{1}{\xi}$, then $D = (\xi - 1)/k^4$, and so the propagator is[54]

[54] The S_ξ *gauges* are a generalization of the Lorenz gauge that is applicable to theories expressed in terms of an action principle with the Lagrangian density $\mathcal{L}$.

$$D_F^{\mu\nu} = \frac{-i}{k^2 + i \in}(g_{\mu\nu} - (1-\xi)\frac{k_\mu k_\nu}{k^2(+i \in)}), \qquad (A.26)$$

If it is physically independent of ξ, that is to say $\xi = 0$, we have the Landau gauge. If $\xi = 1$, then we have the Feynman gauge (primary), since, in general, $\xi = 1$ could be like $m^2 \to 0$ in the Klein-Gordon equation, but $-g^{00} = -1$ casts to A_0, as a negative norm state, which is unphysical [3].

A.3. Methods

We consider our functional derivative

$$\prod_\mu = \frac{\delta \mathcal{L}}{\delta(\partial_0 A^\mu)}, \qquad (A.27)$$

and we want the bracket $\{A_\mu, \prod_\nu\} \approx \eta_{\mu\nu}$. An exercise for our readers is to demonstrate $\prod_0 = 0$!

The gauge invariance implies that A_μ has two physical propagators. Some

---

Instead of *fixing* the gauge by constraining the gauge field *a priori* via an auxiliary equation, one adds a gauge *breaking* term $\delta\mathcal{L} = -\frac{(\partial_\mu A^\mu)}{2\xi}$ to the "physical" (gauge invariant) Lagrangian. The choice of the parameter ξ determines the choice of gauge. The *Landau gauge*, obtained as the limit $\xi \to 0$ is classically equivalent to the Lorenz gauge. Postponing the taking of a limit until after the theory is quantized improves the rigor of certain proofs of existence and equivalence. Most quantum field theory computations are simplest in the *Feynman-'t Hooft gauge*, in which $\xi = 1$; a few are more tractable in other S_ξ gauges, such as the *Yennie gauge* $\xi = 3$.

An equivalent formulation of the S_ξ gauge uses an auxiliary field, a scalar field B with no independent dynamics $\delta\mathcal{L} = B\partial_\mu A^\mu + \frac{\xi}{2}B^2$.

resolutions are given in this respect. For example, the *radiation gauge*, which breaks the covariance. Without sources, we choose the gauge $A_0 = 0, \nabla A = 0.$

Then, selecting the field

$$A_\mu = \int d^3k \frac{1}{\sqrt{2E_k}} \sum_{r=1}^{2} \varepsilon_\mu^r (a_k^r e^{-ikx} + a_k^{r\dagger} e^{ikx}), \qquad (A.28)$$

we have the orthogonal conditions $\varepsilon_0^r = 0, \varepsilon^r \bullet k = 0$ and $\varepsilon^1 \bullet \varepsilon^2 = 0$ (e.g. $k = (\omega, 0, 0, \omega)$, implying that $\varepsilon^1 = (0,1,0,0)$ and $\varepsilon^2 = (0,0,1,0)$.

Then, using the Klein-Gordon equation, we have

$$[a_k^r, (a_{k'}^s)]^\dagger = \delta^{rs} \delta(k - k'), \qquad (A.29)$$

Considering

$$\sum_r \varepsilon_i^r \varepsilon_j^r = \delta_{ij} - \frac{k_i k_j}{|k|^2}, \qquad (A.30)$$

the propagator (A.26) takes the form

$$D_{ij}^r(k) = \frac{i}{k^2 + i\varepsilon} \left(\delta_{ij} - \frac{k_i k_j}{|k|^2} \right), \qquad (A.31)$$

With these sources, we select the gauge $\nabla^2 A^0 = -ej^0$, then D_{00}, including the additional effects. The result is the same as for the $\xi = 1$ gauge [97-99].

Now, we consider other resolutions or approaches, for example, Lorentz invariant quantization (or *Fermi/Gupta/Bleuler* [99, 100])[55]. In this, we consider

[55] In QFT, Gupta-Bleuler formalism is a way of quantizing the electromagnetic field. We consider a single photon. The basis of the 1-photon vector space (which does

$$\mathcal{L} = \frac{1}{2} \partial_\mu A_\nu \partial^\mu A^\nu, \qquad (A.32)$$

then $\partial^2 A_\mu = 0$. Imposing $\partial_\mu A^\mu = 0$, "weakly", we have

$$\langle s | \partial_\mu A^\mu | s \rangle = 0, \qquad (33)$$

Considering expansion of the field

$$A_\mu = \int d^3k \frac{1}{\sqrt{2E_k}} \sum_{r=1}^{3} (\varepsilon_\mu^r a_k^r e^{-ikx} + \varepsilon_\mu^{r*} a_k^{r\dagger} e^{ikx}), \qquad (A.34)$$

with the commutation relations

$$[a_k^r, (a_{k'}^s)^\dagger] = -\eta^{rs} \delta^3(k - k'), \qquad (A.35)$$

we obtain

$$< 0 | a_k^0 a_{k'}^{0\dagger} | 0 >= -\delta^3(k - k'), \qquad (A.36)$$

---

not conform to Hilbert space) is given by the eigenstates $|k, \varepsilon_\mu >,|$ where the 4-momentum is null ($k^2 = 0$), for component k_0 the energy is positive, ε_μ is the unit polarization vector, and the index μ ranges from 0 to 3. So, k is uniquely determined by the spatial momentum $\vec{k}$. Using bracket notation, we equip this space with a sesquilinear form defined by

$$\langle \vec{k}_a; \varepsilon_\mu | \vec{k}_b; \varepsilon_\nu \rangle = (-\eta_{\mu\nu}) \frac{1}{2|\vec{k}_a|} \delta(\vec{k}_a - \vec{k}_b),$$

where the factor $\frac{1}{2|\vec{k}_a|}$ is used to implement Lorentz covariance. Here, we are using $+---$ as the metric signature. However, this sesquilinear form gives positive norms for spatial polarizations, but negative norms for time-like polarizations. Negative probabilities are unphysical, not to mention a physical photon only has two transverse polarizations, not four.

where $a_{k'}^{0\dagger}|0>$ stands for "ghost" states (*the space defined for them does not conform to a Hilbert space*). Then, we have the propagator ($\xi = 1$ gauge)

$$D^{\mu\nu} = \frac{-i}{k^2 + i\varepsilon}\eta^{\mu\nu}, \qquad (A.37)$$

The action given in (A.15) can be rewritten for these resolutions or approaches as the *six-flat-Maxwell field actions*

$$\begin{aligned} S_{\xi} = \int \phi^*[-\partial_{\mu}\partial^{\mu} - m^2]\phi + \\ \frac{1}{2}A_{\mu}[g^{\mu\nu} - (1-\frac{1}{\xi})\partial^{\mu}\partial^{\nu}]A_{\nu} + ieA_{\mu}((\partial^{\mu}\phi^*)\phi - \phi^*\partial^{\mu}\phi) + \\ e^2 A_{\mu}A^{\mu}\phi^*\phi, \qquad (A.38) \end{aligned}$$

We have the following covariant Feynman rules (see Table A.1).

TABLE A.I.: Covariant Feynman Rules

Type of Propagator	Feynman Graph	Propagator
Scalar Propagator	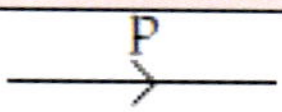	$\dfrac{i}{p^2 - m^2 + i\varepsilon}$
Photon Propagator	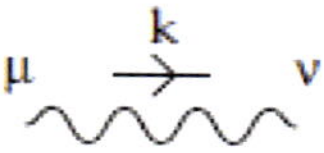	$\dfrac{-i}{k^2 + i\varepsilon}(g_{\mu\nu} - (1-\xi)\dfrac{k_{\mu}k_{\nu}}{k^2})$
$A\phi^*\phi$ vertex	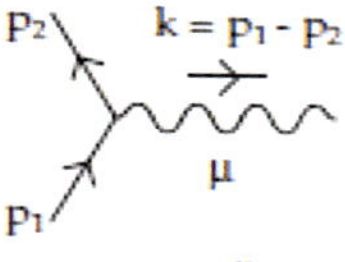	$+ i\varepsilon(p_2 + p_1)^{\mu}$
$A^2\phi^*\phi$ vertex	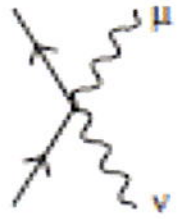	$2ie^2 g^{\mu\nu}$

External edges: one for incoming/outgoing scalar $\varepsilon_{\mu}, \varepsilon_{\mu}^{*}$, for incoming/outgoing photon.

A.4. 6-Flat Maxwell Fields

The invariant $SO(2,4)$-actions on the Minkowskian flat Maxwell fields present a picture of the Maxwell equations in a 6-dimensional frame[56]. The shape of these fields and their orthogonal conditions do not vary.

To obtain 6-Flat Maxwell fields from field A_μ, we apply the *Gupta-Bleuler quantization* scheme mentioned in the footnote of page 34, with the bracket related to the functional derivative given in (A.27). Then, the gauge invariance implies two physical states: one physical state in the Minkowskian frame and other physical states obtained in the Fock-space, including the states obtained in the second quantization due to Gupta-Bleuler. The solutions in Minkowski space can be obtained by

$$A_{lm,\mu}(x) = \varepsilon_\mu(x)\phi_{ml}(x), \qquad \forall x \in \mathbb{M}, \qquad (A.39)$$

where $\varepsilon_\mu(x)$ represents the polarization vectors.

The scalar modes $\phi_{lm}(x)$ are the solutions of the Minkowskian $SO(2,4)$-invariant (or massless) scalar field equation $\partial^2\phi = 0$ (the case of pure energy states of the field equation $\partial^2 A_\mu = 0)$ having

$$\phi_{lm}(x) = c_l Y_{lm}(x), \qquad (A.40)$$

with $Y_{lm}(x)$ and the hyper-spherical harmonics and normalization constant c_l is chosen in order to give (A.29). As a consequence, we have (A.35). However, the Minkowskian modes can be extended via a Weyl transformation [101][57] through

---

56 Using the intersection between an $\mathbb{R}^6$-null-cone with a 5-D surface, we obtain a conformally flat 4-dimensional space. This null-cone is the $SO(2,4)$-invariant.

57 In theoretical physics, a Weyl transformation is a local rescaling of the metric tensor $g_{\mu\nu} \mapsto e^{-2\omega(x)} g_{\mu\nu}$, which produces another metric in the same conformal

$$A'_{lm},_{\mu}(x) = A_{lm,\mu}(x) + \Im_l(x)A_{lm,+}(x), \forall x \in \mathbb{M}, \qquad (A.41)$$

which solves the Maxwell equation in the Eastwood-Singer gauge. The reciprocal transformation obeys the inversion of the Weyl transformation, using the approximation of the path integrals to the structural coefficients.

The propagators for the six-flat Maxwell space are Wightman functions [102]. Other procedures could be used to obtain the Maxwell equations in Minkovskian space from other n-dimensional spaces, although the would need to use hyper-complex analysis tools and chirality. It is necessary to observe other constructions of the m-Flat Maxwell fields through gauge fields.

Scholium 3.1. To obtain m-Flat Maxwell fields, the physical states in n-dimensions with $n>m$ are necessary to solve the Maxwell equations in the frame-gauge corresponding to the second quantization of these n-dimensional physical states with polarization vectors of the field A_{μ}.

---

class. A theory or invariant expression under this transformation is described as conformally invariant, or is said to possess Weyl symmetry. The Weyl symmetry is an important symmetry in conformal field theory.

Appendix B. Measurable Quantities of Quantum Field Theory: A Review of Scalar Field Theory and its Applications

B.1. Introduction

Due to the fact that in those parts of inhomogeneous spacetime, the time modes cannot be separated out, it is usual to use a method with perturbation techniques, based on the *S*-matrix, in which the gravitational non-homogeneities are considered as external tensor fields that are coupled to the quantum fields with a given homogeneous background. One of these techniques uses a gauge particle that breaks with this couple, producing scattering.

Such a gauge particle can be a dilaton or other gauge photon. Through its backreaction with the background, it shows the desired perturbation in spacetime. This can be interpreted as parts of inhomogeneous spacetime and their produced energy can be interpreted as scattering particles. In the context of supersymmetry and using the pair interaction between particles and anti-particles (*for example, electron-positron pair creation or their slicing by partial annihilation, which shows these inhomogeneities*), observables of the field include *quantum curvature and torsion as asymmetries* that require an extension of the Einstein-Cartan theory to one of *fermion self interactions induced by quantum torsion*. In addition, we have the corresponding torsion terms in quantum curvature inside this formal theory [103, 104][58]. In this way, the total action in the Einstein-Cartan theory permitting us to see these interactions is

[58] The method has been extended to conformally flat expanding backgrounds for the case of quantum scalar fields. In this case, if the particles are conformally coupled, the background does not create particles—they are a consequence of the loss of conformal invariance due to the inhomogeneities.

$$S_{\text{TOTAL}} = \frac{1}{2\kappa}\int d^4 x e e_a^{\mu} e_b^{\nu} R_{\mu\nu}^{ab}(\varpi) +$$
$$+\frac{i}{2}\int d^4 x e\left(e_a^{\mu}(\overline{\psi}\tilde{\gamma}^{a}\mathcal{D}_{\mu}(\varpi, A)\psi - \overline{\mathcal{D}_{\mu}(\varpi, A)\psi}\tilde{\gamma}^{a}\psi\right) - \qquad (B.1)$$
$$-\int d^4 x e \frac{3}{10}\kappa J_{(A)}^{\mu} J_{A)\mu},$$

where $\kappa J_{(A)}^{\mu} J_{A)\mu}$ is the fermion self interaction induced by quantum torsion; $J_{(A)}^{\mu}$ is the spinor axial current; ψ and $\overline{\psi}$ are the spinor and the corresponding conjugated form; A is the quantum scalar field; and κ is the gravitational constant.

The following fermion self interaction action induced by quantum torsion

$$S_{\odot} = \int \kappa J_{(A)}^{\mu} J_{A)\mu}, \qquad (B.2)$$

reveals the microscopic aspects of curvature inherent in spin-matter interacting with the electromagnetic fields given by the current terms with gravity.

The action

$$S = \frac{i}{2}\int d^4 x e\left(e_a^{\mu}(\overline{\psi}\tilde{\gamma}^{a}\mathcal{D}_{\mu}(\varpi, A)\psi - \overline{\mathcal{D}_{\mu}(\varpi, A)\psi}\tilde{\gamma}^{a}\psi\right), (B.3)$$

reveals fermionic conditioning between particles that may be observed as the difference in inhomogeneous spacetime.

B.2. The LSZ Reduction Formula

In order to describe scattering experiments, we need to construct appropriate initial and final states, before calculating the scattering amplitude. Realizing the summary of free theory, for one particle state we have

$$|k\rangle = a^{\dagger}(\mathbf{k})|0\rangle, \qquad (B.4)$$

where

$$a^{\dagger}(\mathbf{k}) = -i\int d^3x e^{ikx}\overleftrightarrow{\partial}_0 \varphi(x), \qquad (B.5)$$

The vacuum state is annihilated by all $a's$

$$a(\mathbf{k})|0\rangle = 0, \qquad (B.6)$$

then, one particle state has normalization by[59]

$$\langle k|k'\rangle = (2\pi)^3 2\omega\delta^3(\mathbf{k}-\mathbf{k}'), \qquad (B.7)$$

where this normalization is a Lorentz invariant. Let us define a time-independent operator

$$a_1^{\dagger} = \int d^3k f_1(\mathbf{k}) a^{\dagger}(\mathbf{k}), \qquad (B.8)$$

with a wave packet of width σ

$$f_1(\mathbf{k}) \propto \exp\{-(\mathbf{k}-\mathbf{k}_1)^2/4\sigma^2\}, \qquad (B.9)$$

This creates a particle localized in the momentum space near $\mathbf{k}_1$ and localized in the position space near the origin. To go back to the position space, it is necessary to apply the Fourier transform. *This we leave to our readers as an exercise.*

It is important to remember that $a_1^{\dagger}|0\rangle$ is a state that evolves over time (in terms of the Schrödinger depiction), as the wave packet propagates and spreads out leaving the particle localized far from the origin at $t \to \pm\infty$.

Proceeding with our exposition to two particles $a_1^{\dagger}a_2^{\dagger}|0\rangle$ for $\mathbf{k}_1 \neq \mathbf{k}_2$, this is

---

59 $\omega = +(k^2+m^2)^{1/2}$.

Also, consider that the equality (B.7) is really $[a(\mathbf{k}), a^{\dagger}(\mathbf{k})] = (2\pi)^3 2\omega\delta^3(\mathbf{k}-\mathbf{k}')$.

a state describing two particles that were widely separated in the past.[60]

In the theory of interacting fields, $a_1^\dagger(\mathbf{k})$ is not time independent, that is to say $a_1^\dagger(t)$.

We can guess a suitable initial state as

$$|i\rangle = \lim_{t\to-\infty} a_1^\dagger(t)a_2^\dagger(t)|0\rangle, \qquad (B.10)$$

through which we can normalize the wave packets so that $\langle i|i\rangle = 1$.

Similarly, let us consider a final state

$$|f\rangle = \lim_{t\to+\infty} a_{1'}^\dagger(t)a_{2'}^\dagger(t)|0\rangle \qquad (B.11)$$

Where, again, $\mathbf{k}_{1'} \neq \mathbf{k}_{2'}$ and $\langle f|f\rangle = 1$.

Then, the scattering amplitude is

$$\langle f|i\rangle = \langle 0|a_{1'}(+\infty)a_{2'}(+\infty)a_1^\dagger(-\infty)a_2^\dagger(-\infty)|0\rangle, \qquad (B.12)$$

In this respect, a useful formula is

$$a_1^\dagger(+\infty) - a_1^\dagger(-\infty) = \int_{-\infty}^{+\infty} dt\partial_0 a_1^\dagger(t), \qquad (B.13)$$

where, for the sub-integral expression $\partial_0 a_1^\dagger(t)$, we can consider the state of the momentum

---

60 In the theory of interacting fields $a^\dagger(\mathbf{k})$ is not time independent, which implies $a_1^\dagger(\mathbf{k})$.

$$a^{\dagger}(\mathbf{k}) = -i\int d^3x e^{ikx}\overleftrightarrow{\partial}_0 \varphi(x), \qquad (B.14)$$

The difference $a_1^{\dagger}(+\infty) - a_1^{\dagger}(-\infty)$ takes the form

$$a_1^{\dagger}(+\infty) - a_1^{\dagger}(-\infty) = -i\int d^3k f_1(\mathbf{k})\int d^4x \partial_0 (e^{ikx}\overleftrightarrow{\partial}_0 \varphi(x))$$
$$= -i\int d^3k f_1(\mathbf{k})\int d^4x e^{ikx}(\partial_0^2 + \omega^2)\varphi(x)$$
$$= -i\int d^3k f_1(\mathbf{k})\int d^4x e^{ikx}(\partial_0^2 + \mathbf{k}^2 + m^2)\varphi(x)$$
$$= -i\int d^3k f_1(\mathbf{k})\int d^4x e^{ikx}(\partial_0^2 - \overleftarrow{\nabla}^2 + m^2)\varphi(x)$$
$$= -i\int d^3k f_1(\mathbf{k})\int d^4x e^{ikx}(\partial_0^2 - \overrightarrow{\nabla}^2 + m^2)\varphi(x),$$

In the last integral expression, integration by parts is effected, the surface term is annulled, and the particle is localized (a wave packet is needed). Finally, this integral takes the form

$$-i\int d^3k f_1(\mathbf{k})\int d^4x e^{ikx}(\partial_0^2 - \overrightarrow{\nabla}^2 + m^2)\varphi(x) =$$
$$= -i\int d^3k f_1(\mathbf{k})\int d^4x e^{ikx}(-\partial^2 + m^2)\varphi(x), (B.15)$$

where the term $(-\partial^2 + m^2)\varphi(x)$ is 0 in free theory, but not in an interacting theory. This is that to say, the Lagrangian $\mathcal{L}_1 = \frac{1}{2}g\varphi^3$ implies $(-\partial^2 + m^2)\varphi(x) = \frac{1}{2}g\varphi^2$. Thus, we have

$$a_1^{\dagger}(-\infty) = a_1^{\dagger}(+\infty) + i\int d^3k f_1(\mathbf{k})\int d^4x e^{ikx}(-\partial^2 + m^2)\varphi, \qquad (B.16)$$

or the Hermitian conjugate

$$a_1^{\dagger}(+\infty) = a_1^{\dagger}(-\infty) + i\int d^3k f_1(\mathbf{k})\int d^4x e^{-ikx}(-\partial^2 + m^2)\varphi, \qquad (B.17)$$

The scattering amplitude is

$$\langle f | i \rangle = \langle 0 | \mathrm{T} a_{1'}(+\infty) a_{2'}(+\infty) a_1^\dagger(-\infty) a_2^\dagger(-\infty) | 0 \rangle, \qquad (B.18)$$

where T is the ordering by time (*without changing anything*), generalizing to *n'i* and *n'f* particles

$$\langle f | i \rangle = i^{n+n'} \int d^4 x_1 e^{ik_1 x_1} (-\partial_1^2 + m^2) \cdots$$
$$d^4 x'_1 \, e^{ik'_1 x'_1} (-\partial_{1'}^2 + m^2) \cdots$$
$$\times \langle 0 | \mathrm{T} \varphi(x_1) \cdots \varphi(x'_1) \cdots | 0 \rangle, \qquad (B.19)$$

which is like the *Lehmann-Symanzik-Zimmermann formula* (or LSZ-formula) [105, 106].

We note the initial and final states now have delta-function normalization, with multiparticle generalization of $\langle \mathbf{k} | \mathbf{k}' \rangle = (2\pi)^3 2\omega \delta^3 (\mathbf{k} - \mathbf{k}')$.

We can express the scattering amplitudes in terms of the correlation functions. How then can we calculate the correlation functions in this theory inside QFT?

We have assumed the creation operators of free field theory, which would work comparably in the interacting theory acting on the ground state

$$\langle 0 | \varphi(x) | 0 \rangle = \langle 0 | e^{-iPx} \varphi(0) e^{+iPx} | 0 \rangle = \langle 0 | \varphi(0) | 0 \rangle, \qquad (B.20)$$

where $\langle 0 | \varphi(0) | 0 \rangle$ is a Lorentz invariant number and $P^\mu | 0 \rangle = 0.$ We want

$$\langle 0 | \varphi(0) | 0 \rangle = 0, \qquad (B.21)$$

so that $a_1^\dagger(\pm\infty) | 0 \rangle$ is a single particle state. Otherwise, it would create a linear combination of the ground state and a single particle state. We can always shift the field by a constant $\varphi(x) + \upsilon,$ so that (21).

Also, the one-particle state

$$\langle p|\varphi(x)|0\rangle = \langle p|e^{-iPx}\varphi(0)e^{+iPx}|0\rangle = e^{-ipx}\langle p|\varphi(0)|0\rangle, \qquad (B.22)$$

gives a Lorentz invariant number by $\langle p|\varphi(0)|0\rangle$. We want

$$\langle p|\varphi(0)|0\rangle = 1, \qquad (B.23)$$

since this is in the free field theory $a_1^\dagger(\pm\infty)$, creating a correctly normalized particle state. We can always rescale (*renormalize*) the field by a constant as in (B.23).

What happens in multi-particle states?

We can consider the multi-particle state

$$\begin{aligned} \langle p,n|\varphi(x)|0\rangle &= \langle p,n|e^{-iPx}\varphi(0)e^{+iPx}|0\rangle \\ &= e^{-ipx}\langle p,n|\varphi(0)|0\rangle \\ &= e^{-ipx}A_n(\mathbf{p}), \qquad (B.24) \end{aligned}$$

where, in this case, $A_n(\mathbf{p})$ is a Lorentz invariant number.

In general, $a_1^\dagger(t)$ creates some multi-particle states. One can show that the overlap between a one-particle wave packet and a multi-particle wave packet goes to zero as time goes to infinity [107]. By waiting long enough, we can make the multi-particle contribution to the scattering amplitude as small as we want.

B.3. Path Integrals for Interacting Fields: Correlation Techniques

We consider the following example. Given the following Lagrangian

$$\mathcal{L} = -\frac{1}{2}\partial^\mu\varphi\partial_\mu\varphi - \frac{1}{2}m^2\varphi^2 + \frac{1}{6}g\varphi^3, \qquad (B.25)$$

after shifting and rescaling, we have

$$\mathcal{L} = -\frac{1}{2} Z_{\varphi} \partial^{\mu} \varphi \partial_{\mu} \varphi - \frac{1}{2} Z_m m^2 \varphi^2 + \frac{1}{6} Z_g g \varphi^3 + Y\varphi, (B.26)$$

Indeed, we consider (B.26) as φ*-cubed* QFT-interacting with fields satisfying (B.21) and $\langle k|\varphi(x)|0\rangle = e^{-ikx}$, where $\langle 0|0\rangle = 1$ and $\langle \mathbf{k}'|\mathbf{k}\rangle = (2\pi)^3 2k^0 \delta^3(\mathbf{k} - \mathbf{k}')$.

We wish to evaluate the path integral for this theory as

$$Z(J) = \langle 0|0\rangle_J = \int \mathcal{D}\varphi e^{i\int d^4x\{\mathcal{L}_0 + \mathcal{L}_1 + \mathcal{L}\varphi\}}, \qquad (B.27)$$

which can also be written as

$$Z(J) = e^{i\int d^4x \mathcal{L}_1\left(\frac{1}{i}\frac{\delta}{\delta J(x)}\right)} \int \mathcal{D}\varphi e^{i\int d^4x\{\mathcal{L}_0 + \mathcal{L}\varphi\}}, \qquad (B.28)$$

A proportionality factor can be used to get the correct normalization, that is to say $\propto e^{i\int d^4x \mathcal{L}_1\left(\frac{1}{i}\frac{\delta}{\delta J(x)}\right)} Z(J)$ and always $Z(J) = 1$.

The path integral in the free field theory, we can find as [6]

$$Z_0(J) = \exp\left[\frac{i}{2}\int d^4x d^4x' J(x)\Delta(x - x')J(x')\right], \qquad (B.29)$$

Then, in the case of (B.26), the perturbing Lagrangian is

$$\mathcal{L}_1 = \frac{1}{6} Z_g g \varphi^3 + \mathcal{L}_{ct}, \qquad (B.30)$$

where

$$\mathcal{L}_{ct} = -\frac{1}{2}(Z_{\varphi}-1)\partial^{\mu}\varphi\partial_{\mu}\varphi - \frac{1}{2}(Z_{m}-1)m^{2}\varphi^{2} + Y\varphi, (B.31)$$

is the counterterm Lagrangian. In the limit $g \to 0$, we expect $Y \to 0$ and $Z_i \to 1$ (we find $Y = O(g)$ and $Z_i = 1 + O(g^2)$).

Thus, we get[61]

$$Z(J) = \exp\left[-\frac{1}{2}\int d^4x\left(\frac{1}{i}\frac{\delta}{\delta J(x)}\right)(-A\partial_x^2 + Bm^2)\left(\frac{1}{i}\frac{\delta}{\delta J(x)}\right)\right] \times Z_1(J), \qquad (B.32)$$

In the term $(-A\partial_x^2 + Bm^2)$, we have used integration by parts. This results in a new vertex at which two lines meet, the corresponding vertex factor of the Feynman rule is $(-i)\int d^4x(-A\partial_x^2 + Bm^2)$, and for every diagram with a propagator, there is an additional one with the vertex.

B.4. The *S*-Matrix in the Background Homogeneity of Curved Spacetime

We introduce the interaction picture established by the actions inside (B.27) and (B.28), along with the corresponding *S*-matrix. The *S*-matrix methods on curved backgrounds engage in the analysis of self-interacting and

---

[61] We define $Z_1(J) \propto \exp\left[\frac{i}{\epsilon}Z_g g\int d^4x\left(\frac{1}{i}\frac{\delta}{\delta J(x)}\right)^3\right]Z_0(J),$ in which we can consider $Z_1(0) = 1$. Then the expression the exponential by the infinite series $Z_1(J)$ can take the form

$$Z_1(J) \propto \sum_{V=0}^{\infty}\frac{1}{V!}\left[\frac{i}{6}Z_g g\int d^4x\left(\frac{1}{i}\frac{\delta}{\delta J(x)}\right)^3\right]^V \times$$
$$\times \sum_{P=0}^{\infty}\frac{1}{P!}\left[\frac{i}{2}\int d^4y d^4z J(y)\Delta(y-z)J(z)\right]^P,$$

where V is the vertex and P stands for the propagators. The number of surviving sources (after taking all derivatives into account) E (externally, $E = 2P - 3V$, where $3V$ derivatives can act on $2P$ sources in $2P!/(2P-3V)!$) different ways.

mutually interacting fields in inflationary cosmology and the determination of some observables, such as curvature in the first instance. They can be used to determine the torsion of space-time through pure scalar field theory.

The Lagrangian that describes the interaction of the inhomogeneous gravitational field with the quantum spinor field must be a Lagrangian, as described in (B.30). Indeed, we separate the space-time metric $g_{\mu\nu}$ into an inhomogeneous background part $\widetilde{g}_{\mu\nu}$ and a small inhomogeneous field $\delta g_{\mu\nu}$, as

$$g_{\mu\nu} = \widetilde{g}_{\mu\nu} + \delta g_{\mu\nu}, \qquad (B.33)$$

Then, the Dirac action given by (B.15), in relation to the ambit of the gravitational background and *their spinor formalism*, takes the form[62]

$$S = \int d^4 x e\left[i\overline{\psi}\gamma^{\mu}(x)(\partial_{\mu} + \Gamma_{\mu})\psi - m\overline{\psi}\psi\right], \qquad (B.34)$$

By applying this Dirac action, (34) can be separated into a "free" action and an "interaction" part as

$$S = \overset{0}{S} + \delta S = S + \int d^4 x \left[\frac{\delta S}{\delta g_{\mu\nu}}\right]\delta g_{\mu\nu}, \qquad (B.35)$$

The stress tensor of the spinor field can be deduced from the action as

$$T^{\mu\nu} = \frac{2}{\sqrt{g}}\frac{\delta S}{\delta g_{\mu\nu}}, \qquad (B.36)$$

---

[62] The Dirac action in flat space is

$$S = \int d^4 y\left[\frac{i}{2}\overline{\psi}\gamma^{a}\overleftrightarrow{\partial}_{a}\psi - m\overline{\psi}\psi\right],$$

where $2P$ stands for global inertial coordinates that can easily be expressed in curved space-time using the covariant derivative $D_a = e^{\mu}_{a}\nabla_{\mu}$.

for which we write the interaction term as $\frac{\sqrt{\overset{0}{g}}}{2}\overset{0}{T}^{\mu\nu}\delta g_{\mu\nu}$, e.g. the external inhomogeneous field couples linearly with the stress tensor of the spinor field, as one might expect. The Lagrangian density defined through the action

$$S = \int d^4 x \mathcal{L}, \qquad (B.37)$$

may be written as (*in the same class* $Z_i = 1 + O(g^2)$)

$$\mathcal{L} = \mathcal{L}_0 + \mathcal{L}_1 + O(\delta g^2), \qquad (B.38)$$

having a counterterm Lagrangian as in (B.31). Indeed, having $\mathcal{L}_0$ as the free Lagrangian density and $\mathcal{L}_1$ as the interaction Lagrangian density, we can compute particle production in the frame of perturbation theory, provided that the interaction vanishes in the "in" and "out" regions.

Considering the condition of time given in the state, remember that $a_1^\dagger|0\rangle$ is a state that evolves with time (*in the Schrödinger picture*), as the wave packet propagates and spreads out and so the particle is localized far from the origin at time $\eta \to \pm\infty$ in the interaction

$$\lim_{\eta \to \pm\infty} \delta g_{\mu\nu} = 0, \qquad (B.39)$$

If there is no particle created by the background, the "in" and "out" vacuum can be identified. This is the case for one instant when the background is a flat space, that is to say

$$\overset{0}{g}_{\mu\nu} = \eta_{\mu\nu}, \qquad (B.40)$$

Here, $\eta_{\mu\nu}$ is the Minkowski metric. When it is conformally flat and we consider *conformally-coupled particles, e.g. massless fermions*, we have the same situation.

A physical state given by $|\Psi\rangle$ in interaction involves time according to the Schrödinger equation

$$H_I|\Psi\rangle = i\frac{\partial|\Psi\rangle}{\partial\eta}, \qquad (B.41)$$

where H_I is the interacting Hamiltonian constructed from the interacting Hamiltonian density given by

$$H_I = \int d^3x\mathcal{H}_I, \qquad (B.42)$$

Equation (41) can be formally solved by means of the unique *S*-matrix operator connecting the "in" and "out" states and thus

$$S = \mathrm{T}\exp\{-i\int d^4x\mathcal{H}_I\}, \qquad (B.43)$$

where T is the time-ordering operator (*without changing anything*) as before (*see section 2*). If we start with $|0,in\rangle$, e.g. no "in" particles, the final evolution of this state as a consequence of interaction is

$$|0,out\rangle = S|0,in\rangle, \qquad (B.44)$$

Now, we can expand *S*, "creating perturbation", up to the first order in $\delta g_{\mu\nu}$, which becomes

$$S \cong 1 - i\mathrm{T}\int d^4x\mathcal{H}_I, \qquad (B.45)$$

where

$$\mathcal{H}_I = -\mathcal{L}_I, \qquad (B.46)$$

The Lagrangian has already been described by (B.30).

In our case, however, the interacting Lagrangian contains derivative terms and we must take care. However, using path integral methods it is not hard to see since there is no ordering problem at the quantum level for the Hamiltonian operator (because there is no mixing of canonical conjugate variables in the classical action) and thus the substitution (B.46) still holds. Therefore, we write

$$S = 1 + i\mathrm{T}\int d^4x\mathcal{H}_I \equiv 1 + S^{(1)}, \qquad (B.47)$$

Due to the fact that $\mathcal{L}_I$ is quadratic in the field along with the derivatives, the particles are produced in pairs to the first order. The density amplitude for pair creation is given by the S-matrix element

$$S^{(1)}_{\alpha\beta} \equiv \langle \alpha\beta | S^{(1)} | 0 \rangle = iT\int d^4x \langle \alpha\beta | \mathcal{L}_I | 0 \rangle, \qquad (B.48)$$

where α and β are the quantum numbers necessary to describe the particles. The state $|\alpha\beta\rangle$ indicates two-particles states in the same Fock space as $|0\rangle$. If the background creates particles, these states do not correspond to physical states in the "out" region and the Bogoliubov transformation is needed to relate the two types of states [104]. If the background does not create particles, then $|\alpha\beta\rangle$ corresponds to the physical particle states and the total pair-creation density is

$$W^{(1)} = \sum_{\alpha\beta} \left| S^{(1)}_{\alpha\beta} \right|^2, \qquad (B.49)$$

which has quadratic inhomogeneities. As a result, there is observable energy that can be a field observable.

B.5. Total Action in Cartan-Einstein Theory

The production of fermions in space-time is the first evidence that there is torsion in space-time, as ½-spin particles appear when torsion is manifested. Thus, we present a summary of torsion connections in the table below (B.1).

TABLE B.I.: Summary of Torsion Connections

Theory	Connection	Spinor Current
Dirac Spin-1/2 Fermions in Einstein-Cartan	$\omega_{\mu}^{ab} = \varpi_{\mu}^{ab} + \frac{\kappa}{4} \in_{cd}^{ab} e_{\mu}^{c} J_{(A)}^{d}$	$J_{(A)}^{\mu} \equiv \bar{\psi}\gamma_5\gamma^{\mu}\psi$
N = 1 SUGRA	$\omega_{\mu}^{ab} = \omega_{\mu}^{ab}(e) + \kappa_{\mu}^{ab}(\psi),$ $\kappa_{\mu\alpha\beta}(\psi) = \frac{1}{4}\{\bar{\psi}_{\alpha}\gamma_{\mu}\psi_{\beta} +$ $\bar{\psi}_{\mu}\gamma_{\alpha}\psi_{\beta} - \bar{\psi}_{\mu}\gamma_{\beta}\psi_{\alpha}\}$	$J^{\mu}(\psi) = \frac{1}{2} \in^{\mu\nu\alpha\beta} \bar{\psi}_{\nu}$ $\times\gamma_{\alpha}\psi_{\beta}$
N = 2 SUGRA	$\omega_{\mu}^{ab} = \omega_{\mu}^{ab}(e) + \kappa_{\mu}^{ab}(\psi),$ $\kappa_{\mu\alpha\beta}(\psi) = \frac{1}{4}\{\bar{\psi}_{\alpha}^{I}\gamma_{\mu}\psi_{I\beta} +$ $\bar{\psi}_{\mu}^{I}\gamma_{\alpha}\psi_{I\beta} - \bar{\psi}_{\mu}^{I}\gamma_{\beta}\psi_{I\alpha} + c.c.\}$	$J^{\mu}(\psi) = \frac{1}{2} \in^{\mu\nu\alpha\beta} \bar{\psi}_{\nu}^{I}$ $\times\gamma_{\alpha}\psi_{I\beta}$
N = 4 $SU(4)$ *SUGRA*	$\omega_{\mu ab} = \omega_{\mu ab}(e, \psi, \Lambda)$ $\equiv \omega_{\mu ab}(e) + \kappa_{\mu ab},$ $\kappa_{\mu ab} = \kappa_{\mu ab}(\psi) + \kappa_{\mu ab}(\Lambda),$ $\kappa_{\mu\alpha\beta}(\psi) = \frac{1}{4}\{\bar{\psi}_{\alpha}^{I}\gamma_{\mu}\psi_{I\beta} +$ $\bar{\psi}_{\mu}^{I}\gamma_{\alpha}\psi_{I\beta} - \bar{\psi}_{\mu}^{I}\gamma_{\beta}\psi_{I\alpha} + c.c.\},$ $\kappa_{\mu ab}(\Lambda) = -\frac{1}{4e} \in \mu\alpha\beta\sigma\bar{\Lambda}_{I}\gamma^{\sigma}\Lambda^{I}$	$J^{\mu}(\Lambda) = e\bar{\Lambda}_{I}\gamma^{\mu}\Lambda^{I},$ $J^{\mu}(\psi) = \in^{\mu\nu\alpha\beta} \bar{\psi}_{\nu}^{I} \times$ $\gamma_{\alpha}\psi_{I\beta}$

The inhomogous parts located in curved space-time, have been classified inside the Einstein-Cartan with the Immirzi parameter [109, 110]. The actions as given by (B.37) can be modified by the total derivatives, without affecting the on-shell physics (that is to say the motion equations). However, such terms may play a role in quantum physics. In pure gravity without torsion, one can add the following identically zero term. Due to Bianchi identities in the symmetry properties of the Riemann tensor, we have

$$\in^{\mu\nu\alpha\beta} R_{\mu\nu\alpha\beta}(\varpi) = 0, \qquad (B.50)$$

which is no longer true in the presence of torsion. As such, their action in space-time at least is[63]

---

[63] The *Holst action* [110] is an equivalent formulation of the Palatini action http://en.wikipedia.org/wiki/Palatini_actionfor general relativity in terms of vierbeins (*4-D space-time frame field*) by adding a part of a topological term (Nieh-Yan), which does not alter the classical equations of motion as long as there is no torsion

$$S = \frac{1}{2}\int ee^{\alpha}{}_{I} e^{\beta}{}_{J} (F_{\alpha\beta}{}^{IJ} - \alpha * F_{\alpha\beta}{}^{IJ}) \equiv \frac{1}{2}\int ee^{\alpha}{}_{I} e^{\beta}{}_{J}$$

$$\times (F_{\alpha\beta}{}^{IJ} - \frac{\alpha}{2} \in^{IJ}{}_{KL} F_{\alpha\beta}{}^{KL}),$$

where $e^{\alpha}{}_{I}$ is the tetrad; e is its determinant; $F_{\alpha\beta}{}^{IJ}$ is the curvature considered as a function of the connection $A_{\alpha\beta}{}^{IJ}$; α is a (complex) parameter; and we recover the Palatini action when $\alpha = 0$. This only works in 4-D. To be torsion free means that the covariant derivative defined by the connection $A_{\alpha\beta}{}^{IJ}$, when acting on the Minkowski metric η_{IJ}, vanishes, implying the connection is anti-symmetric in its internal indices *I*, *J*. As with the first order tetradic Palatini action where $e^{\alpha}{}_{I}$ and $A_{\alpha\beta}{}^{IJ}$ are taken as independent variables, variation of the action with respect to the connection $A_{\alpha\beta}{}^{IJ}$ (assuming it to be torsion-free) implies curvature $F_{\alpha\beta}{}^{IJ}$, replaced by the usual (mixed index) curvature tensor $R_{\alpha\beta}{}^{IJ}$, (see article on tetradic Palatini action for definitions). Variation of the first term of the action with respect to the tetrad $e^{\alpha}{}_{I}$, gives the (mixed index) Einstein tensor, while variation of the second term with respect to the tetrad gives a quantity that disappears through the symmetries of the Riemann tensor http://en.wikipedia.org/wiki/Riemann_tensor(specifically the first Bianchi identity); together these imply that Einstein's vacuum field equations hold. The canonical 3+1-Hamiltonian formulation of the Holst action with $\alpha = i$, happens to correspond to

$$S_{Holst} = -\frac{\beta}{4\kappa}\int d^4 x e e_a^{\mu} e_b^{v} \in_{cd}^{ab} Rcd_{\mu v}, \qquad (B.51)$$

which is non-trivial in *torsionful geometries*. In the case of torsion, the motion equations imply

$$T^{\mu} = \frac{3\kappa}{4}\frac{\beta}{\beta^2+1} J_{(A)}^{\mu}, S^{\mu} = \frac{3\kappa}{\beta^2+1} J_{(A)}^{\mu}, q_{\mu v p} = 0, (B.52)$$

Then it is necessary to define an adequate vector T^{μ} with a pseudo-vector $J_{(A)}$. However, with fermions gravity is inconsistent in terms of the mathematics. As such, the Holst term needs to be modifed by the addition of fermions to become a total derivative.

Adding fermion production to the Holst form will be equivalent to putting the additional fermion-piece of the form

$$S_{Holst-fermi} = \frac{+\alpha}{2}\int d^4 x e(\overline{\psi}\gamma^{\mu}\gamma_5 \mathcal{D}_{\mu}(\omega)\psi + \overline{\mathcal{D}_{\mu}(\omega)}\gamma^{\mu}\gamma_5\psi), \qquad (B.53)$$

with $\alpha = const$.

Ashtekar variables, which formulate (complex) general relativity as a special type of Yang-Mills gauge theory. The action was simply seen to be a Palatini action with the curvature tensor replaced by its self-dual part only (see article on self-dual Palatini action).
The canonical 3+1-Hamiltonian formulation of the Holst action for real α was shown to have a configuration variable, which is still a connection, and the theory is still a special kind of Yang-Mills gauge theory, but it has the advantage that it is real, as is the corresponding gauge theory (so we are dealing with real general relativity). This Hamiltonian formulation is the classical starting point of loop quantum gravity (LQG) [110], which imports non-perturbative techniques from lattice gauge theory. The parameter defined by $\beta := 1/\alpha$, is usually referred to as the Barbero-Immirzi parameter [108, 109]. The Holst action finds application in most recent versions of spin foam models, which can be considered path integral versions of LQG.

This fermion action with Dirac kinetic terms gives

$$S_{Dirac-Holst-Fermi} = \frac{i}{2}\int d^4xe\{\overline{\psi}\gamma^{\mu}(1-i\alpha\gamma_5)\mathcal{D}_{\mu}\psi + \\ + \overline{\mathcal{D}_{\mu}\psi}\gamma^{\mu}(1-i\alpha\gamma_5)\psi\}, \qquad (B.54)$$

If $\alpha = \beta$, the Holst total gravity-fermion term becomes the total derivative

$$S_{Holst-total} = -i\frac{\beta}{2}\int d^4x\left[I_{NY} + \partial_{\mu}J^{\mu}_{(A)}\right], \qquad (B.55)$$

In an equivalent theory formulated without Immirzi term to the Einstein-Cartan theory being I_{NY}, this represents the Nieh-Yan topological invariant density, which is a unique Lorentz invariant torsion structure (also known as a Lorentz invariant number in the corresponding torsion state). This Nieh-Yan topological invariant density is given by

$$I_{NY} \equiv e^{\mu\nu\rho\sigma}\left(T^{a}_{\mu\nu}T_{\rho\sigma a} - \frac{1}{2}e^{a}_{\mu}e^{b}_{\nu}R_{\rho\sigma ab}(\omega)\right), \qquad (B.56)$$

Or using

$$\partial_{\mu}T_{\nu\rho\sigma} = \left(T^{a}_{\mu\nu}T_{\rho\sigma a} - \frac{1}{2}e^{a}_{\mu}e^{b}_{\nu}R_{\rho\sigma ab}(\omega)\right), \qquad (B.57)$$

then

$$I_{NY} \equiv \in^{\mu\nu\rho\sigma}\partial_{\mu}T_{\nu\rho\sigma}, \qquad (B.58)$$

Thus, the super(Holst) modifications to Dirac spin-1/2 particle production will be

$$S_{Holst-total} = -i\frac{\beta}{4}\int d^4x\partial_{\mu}J^{\mu}_{(A)} = -\frac{i\beta}{6}\int d^4x \in^{\mu\nu\rho\sigma}\partial_{\mu}T_{\nu\rho\sigma}(\psi), (B59)$$

with current $J^{\mu}_{(A)} \equiv \overline{\psi}\gamma_5\gamma^{\mu}\psi$, which is the spinor axial current.

Also, to $N = 1$ SUGRA, we have

$$S_{Holst-total}[\omega(e,\psi)] = -i\frac{\eta}{4}\int d^4x\partial_\mu J^\mu(\psi)$$
$$= -\frac{i\eta}{2}\int d^4x \in^{\mu\nu\alpha\beta} \partial_\mu T_{\nu\alpha\beta}(\psi), \qquad (B.60)$$

with current $J^\mu(\psi) = \frac{1}{2}\in^{\mu\nu\alpha\beta} \overline{\psi}_\nu\gamma_\alpha\psi_\beta.$

To $N = 2$ SUGRA and $N = 4$ SUGRA, the corresponding actions are

$$S_{Holst-total}[\omega(e,\psi)] = -i\frac{\eta}{4}\int d^4x\partial_\mu J^\mu(\psi), \qquad (B.61)$$

and

$$S_{SHolst4}[\omega(e,\psi,\Lambda)] = -i\frac{\eta}{4}\int d^4x\partial_\mu[J^\mu(\psi) - J^\mu(\Lambda)], (B.62)$$

with corresponding currents

$$J^\mu(\psi) = \in^{\mu\nu\alpha\beta} \overline{\psi}_\nu^I\gamma_\alpha\psi_{I\beta}, \qquad (B.63)$$

and (63) with the current to Λ

$$J^\mu(\psi) = e\overline{\Lambda}_I\gamma^\mu\Lambda^I, \qquad (B.64)$$

to (B.62).

These actions in the super(Holst) modifications allow us to see the field behavior in terms of bosonic strings (see figs. B.1 and B.2) and are used in the Kalb-Ramond re-interpretation of the quantum field[64].

[64] In theoretical physics in general and string theory in particular, the Kalb-Ramond field, also known as the NS-NS B-field, is a quantum field that transforms as a 2-form, that is to say, an antisymmetric tensor field with two indices.

Fig. B.1. The possible total action of the electromagnetic nature realized by a galaxy inside the magnetic configuration of intersidereal rotation fields (movement). The accretion lines produce rays of heavy particles whose ejection results in a current of an attractor to the centre of galaxy. This current is supported by other currents signed by the spinor axial currents.

It generalizes the electromagnetic potential (as has been suggested in some applications in superconducting and fermionic Fock space), but it has two indices instead of one. This difference is related to the fact that the electromagnetic potential is integrated over the one-dimensional worldlines of particles to obtain one contribution to the action, while the Kalb-Ramond field must be integrated over the two-dimensional worldsheet of the string. In particular, while the action for a charged particle moving in an electromagnetic potential is given by

$$-q\int dx^{\mu}A_{\mu},$$

that for a string coupled to the Kalb-Ramond field has the form

$$-\int dx^{\mu}dx^{\nu}B_{\mu\nu},$$

This term in the action implies that the fundamental string of string theory is a source of the NS-NS B-field, much like charged particles are sources of the electromagnetic field. The Kalb-Ramond field appears, together with the metric tensor http://en.wikipedia.org/wiki/Metric_tensorand dilaton, as a set of massless excitations of a closed string.

B.6. Torsion from Photonics and a Possible Axion

Electrogravitational energy produces torsion energy and its detection can be obtained by the *back-reaction on the background* in space-time. A perturbation method considering a Hamiltonian $H = H_1 + \varepsilon H_0$, can help us obtain the curvature. Next, on the base of an action like that given in (B.1), we obtain torsion. Using mirror theory, we can demonstrate the images of these observables as observables in macrospace, considering the equivalences between geometrical and algebraic objects in each case.

This last discussion agrees with the annotations realized by Kalb-Ramond on scalar fields in the early universe, with plenty of them found in extended particle physics models, including: *inflaton, dilaton, axion, and many moduli stacks*. We propose the broson as an axion to bosonic torsion as evidenced through torsionful geometries (see fig. B.2).

Fig. B.2. Torsionful geometry.

Unlike gravitational and electromagnetic fields, which are characterized by a central symmetry, the torsion fields of spinning objects possess axis symmetry. The law of inverse squares is not updated here and so field intensity does not depend on its distance from a source. Also, it possesses a huge penetrative ability in any natural medium. *Law-energetic relict neutrinos* represent the quanta of a torsion field—the "tordions". Torsion fields are, in their nature, similar to gravitational fields, which are unscreenable. The gravitational field, when modeled, is treated as a spinal longitudinal polarization, while the torsion field is treated as the horizontal polarization of a physical vacuum.

Likewise, in string-inspired models and torsionful geometries, for example, matter/anti-matter asymmetry in the universe is translated into *leptogenesis/baryogenesis*. This is because the right-handed neutrinos are not needed for extra *CP*-violations. Also, the *Majorana right-handed neutrino masses* from Kalb-Ramond axions (quantum torsion) go beyond the conventional seesaw and through anomalies.

Recent studies have confirmed that the matter-antimatter asymmetry in the thermal equilibrium of the early universe is due to the background *KR*-torsion. Anomalous Majorana neutrino mass generation exists for the same reason.

B.7. Conclusions

The development of modifications in the global actions of Lagrangians created by scalar field theory has been useful in the conformation of torsionful geometries that are in accord with the TFT, TQFT, and the development of methods of algebraic geometry for isotopic curves and loops of groups in ramified fields that separate gravity and electromagnetism. These can have path integrals defined on those loop groups and other spaces that arise out of the study of the torsion.

Taking into account this study, and considering the strong relation between fermions and torsion, it is only necessary to take the first and second order derivative formalism, which helps us obtain Lorentz invariants.

These help to measure the curvature and torsion as observables of quantum fields when considering the decomposition of the torsion tensor into irreducible components according to their transformation under the local Lorentz group of transformations. In this way, the tensor torsion is

$$T_{\mu\nu\rho} = \frac{1}{3}(T_{\nu} g_{\mu\rho} - T_{\rho} g_{\mu\nu}) - \frac{1}{6} \in_{\mu\nu\rho\sigma} S^{\sigma} + q_{\mu\nu\rho}, (B.65)$$

with evidence of the trace

$$T_{\mu} = T^{\nu}_{\mu\nu}, \qquad (B.66)$$

the pseudotrace axial vector

$$S_\mu = \in_{\mu\nu\rho\sigma} T^{\nu\rho\sigma}, \qquad (B.67)$$

and the antisymmetric tensor

$$q^{\nu}_{\rho\nu} = 0 = \in^{\sigma\mu\nu\rho} q_{\mu\nu\rho}, \qquad (B.68)$$

Then, the global action of gravity is

$$S_G = \frac{1}{2\kappa}\int d^4 x e(e^{\mu}_{a} e^{\nu}_{a} R^{ab}{}_{\mu\nu}(\varpi) + \frac{1}{24} S_\mu S^\mu - \frac{2}{3} T_\mu T^\mu + \\ + \frac{1}{2} q_{\mu\nu\rho} q^{\mu\nu\rho}), \qquad (B.69)$$

Using the Dirac matrices

$$\widetilde{\gamma}^a \widetilde{\gamma}^b \widetilde{\gamma}^c = \eta^{ab} \widetilde{\gamma}^c + \eta^{bc} \widetilde{\gamma}^a - \eta^{ac} \widetilde{\gamma}^b - i \in^{abcd} \widetilde{\gamma}_5 \widetilde{\gamma}_d, (B.70)$$

where $\widetilde{\gamma}_5 = i\widetilde{\gamma}^0 \widetilde{\gamma}^1 \widetilde{\gamma}^2 \widetilde{\gamma}^3$ for the fermionic part of the action, we obtain

$$S_{QED-fermions} = \frac{i}{2}\int d^4 x e(e^{\mu}_{a} (\overline{\psi} \widetilde{\gamma}^a \mathcal{D}_\mu(\varpi, A)\psi - \\ - \overline{\mathcal{D}_\mu(\varpi, A)\psi} \widetilde{\gamma}^a \psi) + \frac{i}{4} S_\mu J^{\mu}_{(A)}), \qquad (B.71)$$

in which the term $\frac{i}{4} S_\mu J^{\mu}_{(A)}$ is analogous to the terms in (B.70) given by $\frac{1}{24} S_\mu S^\mu, \frac{2}{3} T_\mu T^\mu$ and $\frac{1}{2} q_{\mu\nu\rho} q^{\mu\nu\rho}$.

The variation of the combined actions $S_G + S_{Fermion}$, written through the *S*, *T*, and *q* components of torsion viewed as independent fields, yields

$$T^\mu = 0, \qquad \frac{1}{24\kappa} S^\mu - \frac{1}{8} J^{\mu}_{(A)} = 0, \qquad q_{\mu\nu\rho} = 0, \qquad (B.72)$$

With the spinor axial current $J^{\mu}_{(A)} \equiv \overline{\psi}\gamma_5\gamma^{\mu}\psi$.

These solutions allow for the spin connection with torsion to be written as Dirac spin-1/2 fermions in Einstein-Cartan theory. However, according to the conjecture given in section 4, the fermionic field theories with **H**-torsion, that is to say, the torsion obtained by the perturbation method considering a Hamiltonian $H = H_1 + \varepsilon H_0$, let us see the effective contribution of its action after integrating out quantum torsion fluctuations and, for the fermions, with S_{ψ} such that

$$-\frac{3}{4}\int d^4\sqrt{-g}S_{\mu}\overline{\psi}\gamma^{\mu}\gamma_5\psi = -\frac{3}{4}\int S \wedge {}^*J^5 + \\ + \text{standard Dirac terms without torsion}, \qquad (B.73)$$

Conservation at the quantum level by adding counterterms is satisfied and the scalar field theory proceeds in this frame of study. One possibly way can be the consideration of dynamic pseudo-scalar fields.

Appendix C. Path Integrals for Interacting Fields

C.1. Canonical Introduction

Realizing the summary of free theory (realized in JPS-Mayo 14), to one particle state, we have

$$|k\rangle = b^{\dagger}(\mathbf{k})|0\rangle, \qquad (C.1)$$

where

$$b^{\dagger}(\mathbf{k}) = -i\int d^3x e^{ikx}\overleftrightarrow{\partial}_0\varphi(x), \qquad (C.2)$$

The vacuum state is annihilated by all $b's$

$$b(\mathbf{k})|0\rangle = 0, \qquad (C.3)$$

and thus one particle state has normalization given by[65]

$$\langle k|k'\rangle = (2\pi)^3 2\omega\delta^3(\mathbf{k}-\mathbf{k}'), \qquad (C.4)$$

where this normalization is a Lorentz invariant.

Let us define a time-independent operator

$$b_1^{\dagger} = \int d^3k f_1(\mathbf{k}) b^{\dagger}(\mathbf{k}), \qquad (C.5)$$

with a wave packet of width σ

$$f_1(\mathbf{k}) \propto \exp\{-(\mathbf{k}-\mathbf{k}_1)^2/4\sigma^2\}, \qquad (C.6)$$

This creates a particle localized in the momentum space near $\mathbf{k}_1$ and

---

65 $\omega = +(k^2+m^2)^{1/2}$.

Also consider that the equality (C.7) is, in reality $[b(\mathbf{k}), b^{\dagger}(\mathbf{k})] = (2\pi)^3 2\omega\delta^3(\mathbf{k}-\mathbf{k}')$.

localized in the space position near the origin. To go back to the space position, it is necessary to apply the Fourier transform.

It is also necessary remember that $b_1^\dagger|0\rangle$ is a state that evolves over time (in the Schrödinger picture), as the wave packet propagates and spreads out. Thus, the particle is localized far from the origin at $t \to \pm\infty$.

Proceeding with our exposition to two particles, $b_1^\dagger b_2^\dagger|0\rangle$ for $\mathbf{k}_1 \neq \mathbf{k}_2$ is a state describing two particles widely separated in the past[66].

In the interacting field theory, $b_1^\dagger(\mathbf{k})$ is not time independent, that is to say $b_1^\dagger(t)$.

A guess of a suitable initial state is

$$|i\rangle = \lim_{t\to-\infty} b_1^\dagger(t) b_2^\dagger(t)|0\rangle, \qquad (C.7)$$

for which we can normalize the wave packets, so that $\langle i|i\rangle = 1$.

Similarly, let us consider the final state

$$|f\rangle = \lim_{t\to+\infty} b_{1'}^\dagger(t) b_{2'}^\dagger(t)|0\rangle, \qquad (C.8)$$

where, again, $\mathbf{k}_{1'} \neq \mathbf{k}_{2'}$ and $\langle f|f\rangle = 1$.

Then, the scattering amplitude is

$$\langle f|i\rangle = \langle 0|b_{1'}(+\infty) b_{2'}(+\infty) b_1^\dagger(-\infty) b_2^\dagger(-\infty)|0\rangle, \qquad (C.9)$$

In this respect, a useful formula is

66 In the interacting field theory, $b^\dagger(\mathbf{k})$ is not time independent, which implies $b_1^\dagger(\mathbf{k})$.

$$b_1^\dagger(+\infty) - b_1^\dagger(-\infty) = \int_{-\infty}^{+\infty} dt \partial_0 b_1^\dagger(t), \qquad (C.10)$$

where, for the sub-integral expression $\partial_0 b_1^\dagger(t)$, we can consider the state of the momentum

$$b^\dagger(\mathbf{k}) = -i\int d^3x e^{ikx} \overleftrightarrow{\partial}_0 \varphi(x), \qquad (C.11)$$

Thus, the difference $b_1^\dagger(+\infty) - b_1^\dagger(-\infty)$ takes the form

$$\begin{aligned}
b_1^\dagger(+\infty) - b_1^\dagger(-\infty) &= -i\int d^3k f_1(\mathbf{k}) \int d^4x \partial_0 (e^{ikx} \overleftrightarrow{\partial}_0 \varphi(x)) \\
&= -i\int d^3k f_1(\mathbf{k}) \int d^4x e^{ikx} (\partial_0^2 + \omega^2) \varphi(x) \\
&= -i\int d^3k f_1(\mathbf{k}) \int d^4x e^{ikx} (\partial_0^2 + \mathbf{k}^2 + m^2) \varphi(x) \\
&= -i\int d^3k f_1(\mathbf{k}) \int d^4x e^{ikx} (\partial_0^2 - \overleftarrow{\nabla}^2 + m^2) \varphi(x) \\
&= -i\int d^3k f_1(\mathbf{k}) \int d^4x e^{ikx} (\partial_0^2 - \overrightarrow{\nabla}^2 + m^2) \varphi(x),
\end{aligned}$$

where in the last integral expression integration by parts is effected and the surface term is annulled and the particle is localized (the wave packet is needed). Finally, this integral takes the form

$$\begin{aligned}
&-i\int d^3k f_1(\mathbf{k}) \int d^4x e^{ikx} (\partial_0^2 - \overrightarrow{\nabla}^2 + m^2) \varphi(x)\Big|_b = \\
&\qquad = -i\int d^3k f_1(\mathbf{k}) \int d^4x e^{ikx} (-\partial^2 + m^2) \varphi(x), \qquad (C.12)
\end{aligned}$$

where the term $(-\partial^2 + m^2)\varphi(x)$ is 0 in free theory, but not in interacting field theory. That is to say, the Lagrangian $\mathcal{L}_1 = \frac{1}{2} g \varphi^3$ implies $(-\partial^2 + m^2)\varphi(x) = \frac{1}{2} g \varphi^2$.

Thus, we have

$$b_1^\dagger(-\infty) = b_1^\dagger(+\infty) + i\int d^3k f_1(\mathbf{k})\int d^4x e^{ikx}(-\partial^2 + m^2)\varphi, \qquad (C.13)$$

or the Hermitian conjugate

$$b_1^\dagger(+\infty) = b_1^\dagger(-\infty) + i\int d^3k f_1(\mathbf{k})\int d^4x e^{-ikx}(-\partial^2 + m^2)\varphi, \qquad (C.14)$$

The scattering amplitude is

$$\langle f|i\rangle = \langle 0|\mathrm{T}b_{1'}(+\infty)b_{2'}(+\infty)b_1^\dagger(-\infty)b_2^\dagger(-\infty)|0\rangle, \qquad (C.15)$$

where T is the time ordering (*without changing anything*), generalizing to the $n'i$ and $n'f$ particles

$$\begin{aligned}\langle f|i\rangle_b &= i^{n+n'}\int d^4x_1 e^{ik_1x_1}(-\partial_1^2 + m^2)\cdots \\ &d^4x'_1\, e^{ik'_1x'_1}(-\partial_{1'}^2 + m^2)\cdots \\ &\times\langle 0|\mathrm{T}\varphi(x_1)\cdots\varphi(x'_1)\cdots|0\rangle, \qquad (C.16)\end{aligned}$$

which is the *Lehmann-Symanzik-Zimmermann formula* (or LSZ-formula) [3, 4].

We note that the initial and final states now have delta-function normalization and multi-particle generalization of $\langle \mathbf{k}|\mathbf{k}'\rangle_b = (2\pi)^3 2\omega\delta^3(\mathbf{k} - \mathbf{k}')$.

Then, we can express the scattering amplitudes in terms of correlation functions. How can we calculate the correlation functions in this theory in QFT?

We have assumed the creation operators of free field theory, which would work comparably in the interacting field theory acting on the ground state

$$\langle 0|\varphi(x)|0\rangle = \langle 0|e^{-iPx}\varphi(0)e^{+iPx}|0\rangle = \langle 0|\varphi(0)|0\rangle, \qquad (C.17)$$

where $\langle 0|\varphi(0)|0\rangle$ is a Lorentz invariant number and $P^\mu|0\rangle = 0.$ We want

$$\left\langle 0 \middle| \varphi(0) \middle| 0 \right\rangle_b = 0, \qquad (C.18)$$

so that $b_1^\dagger(\pm\infty)\left|0\right\rangle$ is a single particle state. Otherwise, it would create a linear combination of the ground state and a single particle state. We can always shift the field by the constant $\varphi(x)+\upsilon,$ as in (C.18).

The one-particle state

$$\left\langle p \middle| \varphi(x) \middle| 0 \right\rangle_b = \left\langle p \middle| e^{-iPx}\varphi(0)e^{+iPx} \middle| 0 \right\rangle = e^{-ipx}\left\langle p \middle| \varphi(0) \middle| 0 \right\rangle, \qquad (C.19)$$

gives a Lorentz invariant number by $\left\langle p \middle| \varphi(0) \middle| 0 \right\rangle$. Likewise, we want

$$\left\langle p \middle| \varphi(0) \middle| 0 \right\rangle_b = 1, \qquad (C.20)$$

In free field theory, $b_1^\dagger(\pm\infty)$ creates a correctly normalized particle state. We can always rescale (*renormalize*) the field by a constant to give (C.20).

What happens in multi-particle states?

Here, we consider the multi-particle state

$$\begin{aligned} \left\langle p,n \middle| \varphi(x) \middle| 0 \right\rangle_b &= \left\langle p,n \middle| e^{-iPx}\varphi(0)e^{+iPx} \middle| 0 \right\rangle \\ &= e^{-ipx}\left\langle p,n \middle| \varphi(0) \middle| 0 \right\rangle = e^{-ipx}A_n(\mathbf{p}), \end{aligned} \qquad (C.21)$$

where $A_n(\mathbf{p})$ is an invariant Lorentz number.

In general, $b_1^\dagger(t)$ creates some multi-particle states. We can show that the overlap between a one-particle wave packet and a multi-particle wave packet goes to zero when the time goes to infinity [5]. A long enough time, we can make the multi-particle contribution to the scattering amplitude as small as we want.

B.2. Path Integral for Interacting Fields: Specific Rules

We consider the scalar field theory defined by the Lagrangian and considering an interacting φ^3 -QFT

$$\mathcal{L} = -\frac{1}{2} Z_\varphi \partial^\mu \varphi \partial_\mu \varphi - \frac{1}{2} Z_m m^2 \varphi^2 + \frac{1}{6} Z_g g \varphi^3 + Y\varphi, \qquad (C.22)$$

with fields satisfying

$$\langle 0 | \varphi(x) | 0 \rangle = 0, \qquad \langle k | \varphi(x) | 0 \rangle = e^{-ikx}, \qquad (C.23)$$

remembering that $\langle 0|0\rangle = 1$ and the normalizing rule $\langle k'|k\rangle = (2\pi)^3 2k^0 \delta^3(\mathbf{k'} - \mathbf{k})$ are always satisfied. We wish to evaluate the *path integral* for this theory

$$Z(J) \equiv \langle 0|0\rangle_J = \int \mathcal{D}\varphi e^{i\int d^4x(\mathcal{L}_0 + \mathcal{L}_1 + J\varphi)} \qquad (C.24)$$

Considering certain properties as functional derivative in respect to $J(x)$ in (C.24), we can re-write the integral as

$$Z(J) = e^{i\int d^4x \mathcal{L}_1\left(\frac{1}{i}\frac{\delta}{\delta J(x)}\right)} \int \mathcal{D}\varphi e^{i\int d^4x(\mathcal{L}_0 + J\varphi)} \propto e^{i\int d^4x \mathcal{L}_1\left(\frac{1}{i}\frac{\delta}{\delta J(x)}\right)} Z_0(J),$$
$$(C.25)$$

where the proportionality sign $\propto$ leads to an additional factor. To get the correct normalization, we require $Z(0) = 1$. Then, for the path integral of free field theory, we find

$$Z_0(J) = \exp\left[\frac{i}{2}\int d^4x d^4x' J(x)\Delta(x-x')J(x')\right], \qquad (C.26)$$

In (C.25), it is assumed that

$$\mathcal{L}_0 = -\frac{1}{2} Z_\varphi \partial^\mu \varphi \partial_\mu \varphi - \frac{1}{2} Z_m m^2 \varphi^2, \qquad (C.27)$$

Thus, in the case of (C.22), the perturbating Lagrangian is

$$\mathcal{L}_1 = \frac{1}{6} Z_g g \varphi^3 + \mathcal{L}_{\text{ct}}, \qquad (C.28)$$

where

$$\mathcal{L}_{\text{ct}} = -\frac{1}{2}(Z_\varphi - 1)\partial^\mu \varphi \partial_\mu \varphi - \frac{1}{2}(Z_m - 1) m^2 \varphi^2 + Y\varphi, \qquad (C.29)$$

which is the counterterm Lagrangian.

In the limit $g \to 0,$ we expect $Y \to 0$ and $Z_i \to 1$. We find $Y = O(g)$ and $Z_i = 1 + O(g^2)$.

Let us look at $Z(J)$ (ignoring the counterterm for now). We define

$$Z_1(J) \propto \exp\left[\frac{i}{6} Z_g g \int d^4x \left(\frac{1}{i}\frac{\delta}{\delta J(x)}\right)^3\right] Z_0(J), \qquad (C.30)$$

Using the expansion series to the exponential, (C.30) takes the form

$$Z_1(J) \propto \sum_{V=0}^{\infty} \frac{1}{V!}\left[\frac{iZ_g g}{6}\int d^4x \left(\frac{1}{i}\frac{\delta}{\delta J(x)}\right)^3\right]^V \times$$
$$\times \sum_{P=0}^{\infty} \frac{1}{P!}\left[\frac{i}{2}\int d^4y d^4z J(y)\Delta(y-z)J(z)\right]^P, \qquad (C.31)$$

Let us look at a term with particular values of P (propagators) and V (vertices):

"*The number of surviving sources (after taking all derivatives) E (for external) is*

$$E = 2P - 3V, \qquad (C.32)$$

Then, $3V$ *derivatives can act on* $2P$ *sources in* $(2P)!/(2P-3V)!$, *different ways*". That is to say, for $V=2$ and $P=3$ we have 6! different terms. For example, if $V=2$, $E=0$, ($P=3$), the two summands in (C.31) give the following

$$Summands = \frac{1}{2!}\left[\frac{iZ_g g}{6}\int d^4x\left(\frac{1}{i}\frac{\delta}{\delta J(x)}\right)^3\right] \times\left[\frac{iZ_g g}{6}\int d^4x\left(\frac{1}{i}\frac{\delta}{\delta J(x)}\right)^3\right]\times$$

$$\times\frac{1}{3!}\left[\frac{i}{2}\int d^4y_1 d^4z_1 J(y_1)\Delta(y_1-z_1)J(z_1)\right]\times$$

$$\times\left[\frac{i}{2}\int d^4y_2 d^4z_2 J(y_2)\Delta(y_2-z_2)J(z_2)\right]\times$$

$$\times\left[\frac{i}{2}\int d^4y_3 d^4z_3 J(y_3)\Delta(y_3-z_3)J(z_3)\right], \qquad (C.33)$$

which establishes the numerical continued product for the corresponding Trotter product of rhte integrals

$$summands = \frac{3!\ 3!\ 2\ 2\ 2}{2!\ 6\ 6\ 3!\ 2\ 2\ 2}\exp\{\int dx_1$$

$$\int dx_2 (iZ_g g)^2\frac{1}{i}\Delta(x_1-x_2)\frac{1}{i}\Delta(x_1-x_2)\times\frac{1}{i}\Delta(x_1-x_2)\}$$

$$=\frac{1}{12}\exp\{\int dx_1\int dx_2(iZ_g g)^2\frac{1}{i}\Delta(x_1-x_2)\frac{1}{i}\Delta(x_1-x_2)\times$$

$$\times\frac{1}{i}\Delta(x_1-x_2)\} \qquad (C.34)$$

However, this integral (C.34) must be corrected to the symmetry factor[67], as predicted from (C.25), which, for our case is $s = 3!\ 2$. Given that this symmetry is paired, we then multiply by ½.

[67] Symmetry factors are related to the symmetries of the Feynman diagram.

$$Z_1(J)=\frac{1}{24}\exp\{\int dx_1\int dx_2(iZ_g g)^2\frac{1}{i}\Delta(x_1-x_2)\times$$
$$\frac{1}{i}\Delta(x_1-x_2)\times\frac{1}{i}\Delta(x_1-x_2)\}, \qquad (C.35)$$

If our sources have the symmetry factor $s=2^3$, then their parity is 1. Thus, in this case the summand product in (C.33) takes the form

$$Z_1(J)=\frac{3!\ 3!\ 3!\ 2}{2!\ 6\ 6\ 3!\ 2\ 2\ 2}\exp\{\int dx_1$$
$$\int dx_2(iZ_g g)^2\frac{1}{i}\Delta(x_1-x_2)\frac{1}{i}\Delta(x_1-x_2)\times\frac{1}{i}\Delta(x_1-x_2)\}$$
$$=\frac{1}{8}\exp\{\int dx_1\int dx_2(iZ_g g)^2\frac{1}{i}\Delta(x_1-x_2)\frac{1}{i}\Delta(x_1-x_2)\times$$
$$\times\frac{1}{i}\Delta(x_1-x_2)\}, \qquad (C.36)$$

A) B)

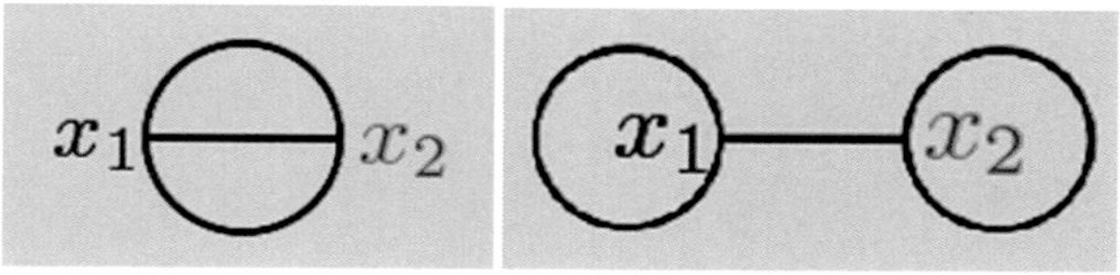

Fig C.1. All connected diagrams with $E=0$ and $V=2$. The two endpoints (B) of each propagator can be swapped and the effect is duplicated by swapping the two vertices. The propagators can be rearranged in 3! ways and all these rearrangements can be duplicated by exchanging the derivatives at the vertices.

We classify the Feynman diagrams in table C.1.

TABLE C.I.: Feynman Diagrams

Graph Element	Functional Element and their Functional Representation	Graph Representation
Line Segment	Propagator $\frac{1}{i}\Delta(x-y)$	
Vertex joining three line segments	Junction $iZ_g g\int d^4x$	
Filled circle at one end of a line segment	Source $i\int d^4xJ(x)$	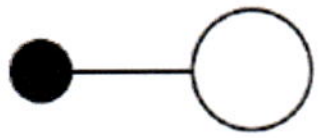

We have one remark about symmetry factors concerning the over-counting of the number of terms giving the same result. This happens when a rearrangement of the derivatives give the same match-up to the sources as a rearrangements of the sources. This is always connected to some symmetry property of the diagram. The factor that we over-counted was the symmetry factor.

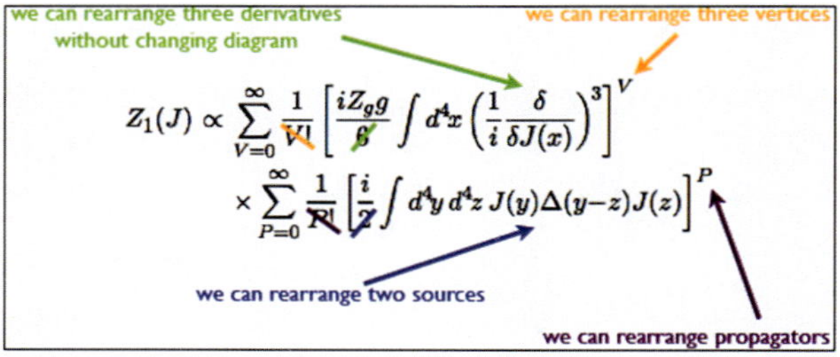

Fig. C.2. Identification of Feynman diagram elements with the analytical expression given in (31).

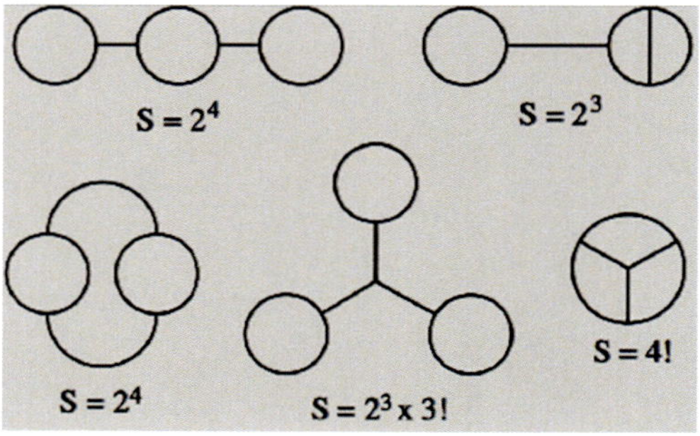

Fig. C.3. All connected diagrams with $E = 0$ and $V = 4$.

A)
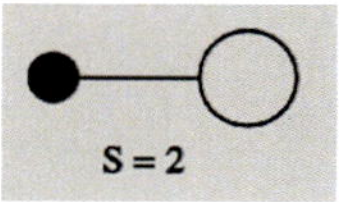

B)
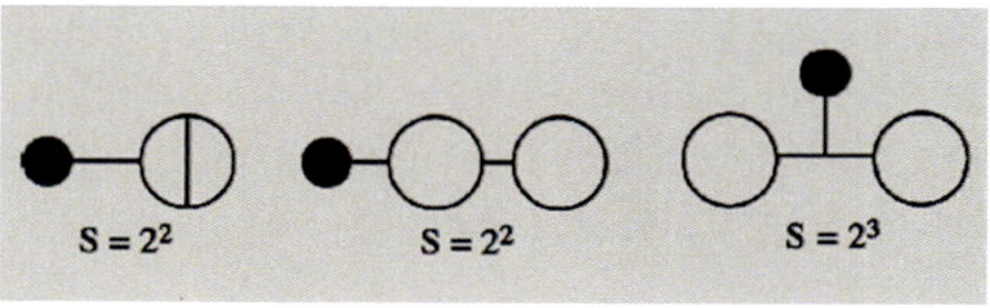

Fig. C.4. A) All connected diagrams with $E = 1$ and $V = 1$. B) All connected diagrams with $E = 1$ and $V = 3$.

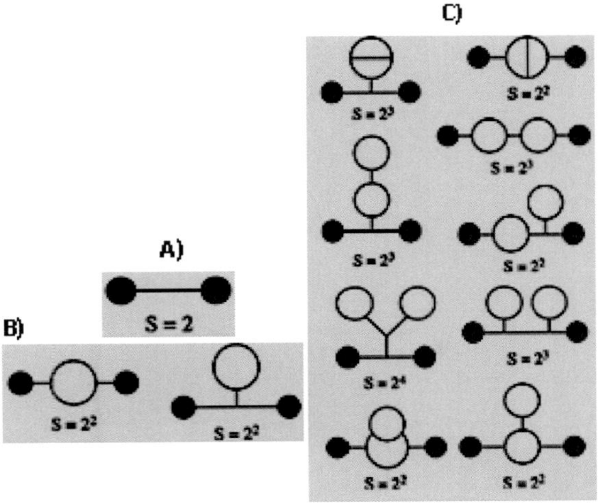

Fig. C.5. A) All connected diagrams with $E = 2$ and $V = 0$. B) All connected diagrams with $E = 2$ and $V = 2$. C) All connected diagrams with $E = 2$ and $V = 4$.

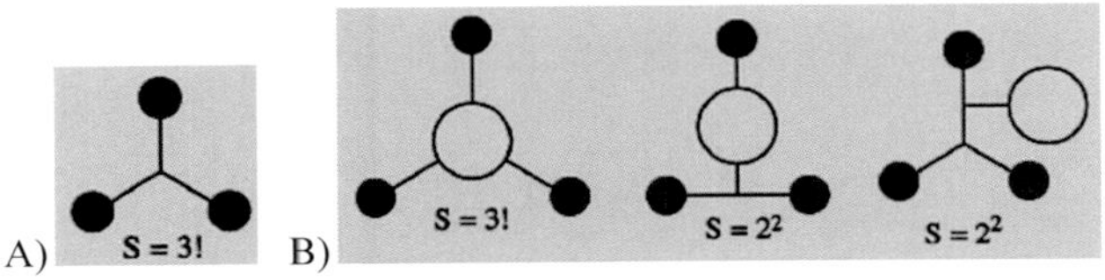

Fig. C.6. A) All connected diagrams with $E = 3$ and $V = 1$. B) All connected diagrams with $E = 3$ and $V = 3$.

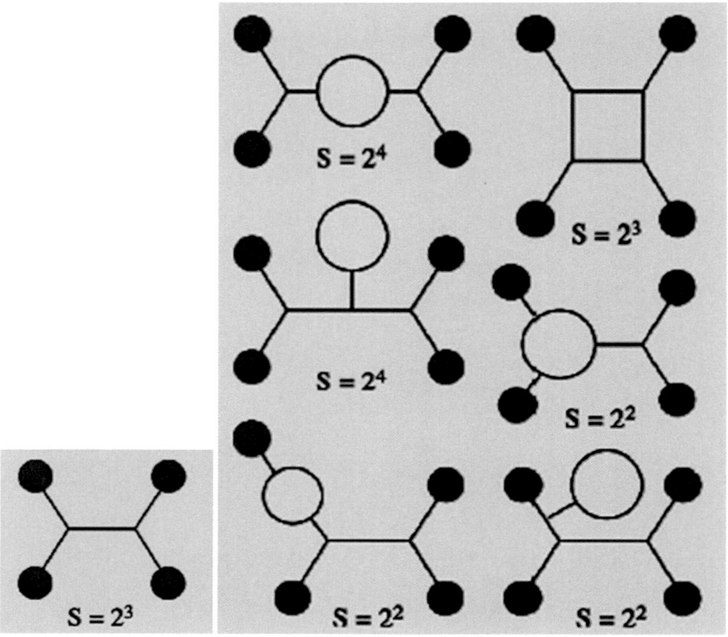

A) B)

Fig. C.7. A) All connected diagrams with $E = 4$ and $V = 2$. B) All connected diagrams with $E = 4$ and $V = 4$.

All these diagrams (figs. C.1, C.3, C.4, C.5, C.6, & C.7) are connected, but $Z(J)$ also contains diagrams that are products of several connected diagrams. For example, for $V = 4$ and $E = 0$ (with $P = 6$); in addition to the connected diagrams we also have figure C.8.

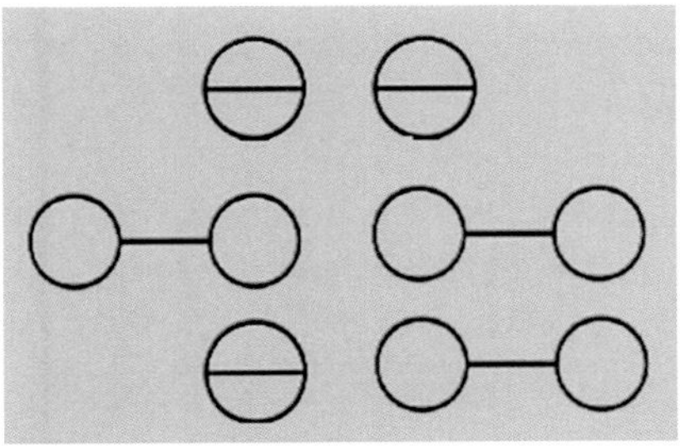

Fig. C.8. Several product types can be realized.

A general diagram D can be written as

$$D = \frac{1}{S_D} \prod_I (C_I)^{n_I}, \qquad (C.37)$$

where S_D comes as

$$S_D = \prod_I n_I!, \qquad (C.38)$$

and this is the additional symmetry factor not already accounted for in the symmetry factors of the connected diagrams (*it is nontrivial only if D contains identical* $C_I^{'s}$) n_I is the number of given $C_I^{'s}$ in *D*. C_I is the particular connected graphs.

$Z_1(J)$ is given by summing all diagrams D

$$\begin{aligned} Z_1(J) &\propto \sum_{(n_I)} D \\ &\propto \sum_{(n_I)} \prod_I \frac{1}{n_I!} (C_I)^{n_I} \\ &\propto \prod_I \sum_{n_i=0}^{\infty} \frac{1}{n_I!} (C_I)^{n_I} \\ &\propto \prod_I \exp(C_I) \propto \exp(\sum_I C_I), \end{aligned}$$

Thus we have found that $Z_1(J)$ is given by the exponential of the sum of the connected diagrams.

Imposing the normalization $Z_1(0)=1$ means we can omit a vacuum diagram (those with no sources) and thus we have

$$Z_1(J) = -\exp[iW_1(J)], \qquad iW_1(J) \equiv \sum_{I\neq\{0\}} C_I, \qquad (C.40)$$

Here, $W_1(0)=0$.

If there are no counterterms, we have

$$Z(J) = Z_1(J),$$

In this case, the vacuum expectation value of the field is

$$\langle 0|\varphi(x)|0\rangle = \frac{1}{i}\frac{\delta}{\delta J(x)} Z_1(J)\Big|_{J=0}$$

$$= \frac{\delta}{\delta J(x)} W_I(J)\Big|_{J=0}, \qquad (C.41)$$

Only diagrams with one source contribute, that is to say, diagrams of this type (see fig. C.9).

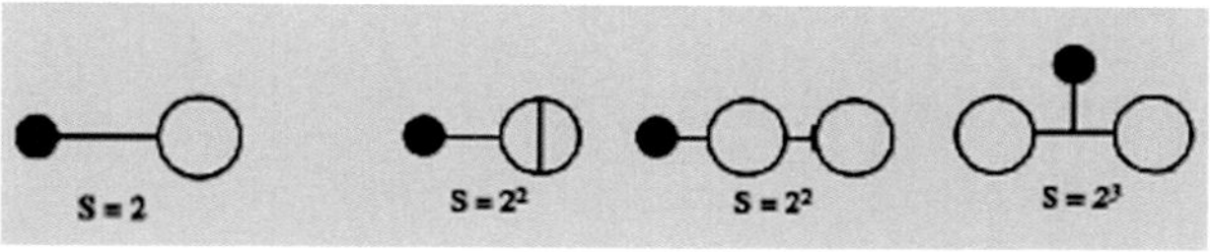

Fig. C.9. Diagrams with one source. In the last three diagrams the source is removed by the derivative.

Then we find[68]

---

68 We use $Z_g=1$ since we know $Z_g = 1 + O(g^2)$.

$$\langle 0|\varphi(x)|0\rangle = \frac{1}{2} ig \int d^4x \frac{1}{i}\Delta(x-y)\frac{1}{i}\Delta(x-y) + O(g^3), \qquad (C.42)$$

which is not zero, as required for the LSZ, and so we need a counterterm.

Including the term $Y\varphi$ in the interaction Lagrangian results in a new type of vertex on which a line segment ends (see fig. C.10).

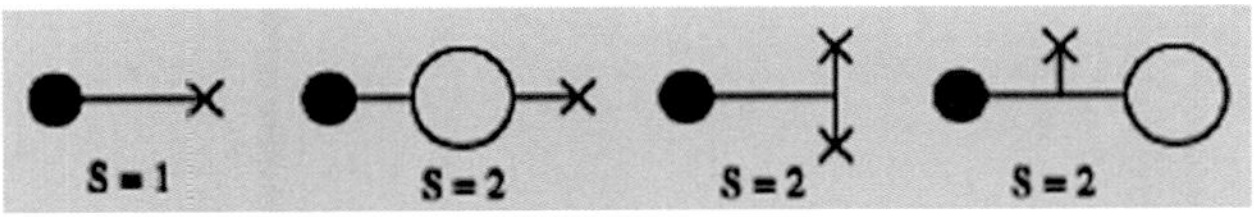

Fig. C.10. Diagrams with one line segment ends.

The corresponding Feynman rule is $iY\int d^4x.$ At the lowest order of g only  contributes

$$\langle 0|\varphi(x)|0\rangle = (iY + \frac{1}{2}(ig)\frac{1}{i}\Delta(0))\int d^4y \frac{1}{i}\Delta(x-y) +$$
$$+ O(g^3), \qquad (C.44)$$

However, in order to satisfy $\langle 0|\varphi(x)|0\rangle = 0$, we have to choose

$$Y = \frac{1}{2} ig\Delta(0) + O(g^3), \qquad (C.45)$$

where $\Delta(0)$ is given by

$$\Delta(C) = \int \frac{d^4k}{(2\pi)^4} \frac{1}{k^2 + m^2 - i\in}, \qquad (C.46)$$

Note that $\Delta(0)$ must be purely imaginary so that Y is real and, in addition, the integral over k is ultraviolet divergent.

To make sense of this, we introduce an ultraviolet cut-off for $\Lambda \approx m$. In order to keep the Lorentz transformation properties of the propagator, we make the following replacement

$$\Delta(x-y) \to \int \frac{d^4k}{(2\pi)^4} \frac{e^{ik(x-y)}}{k^2+m^2-i\in}\left(\frac{\Lambda^2}{k^2+\Lambda^2-i\in}\right), \qquad (C.47)$$

The integral is now convergent, as

$$\Delta(0) = \frac{i}{16\pi^2}\Lambda^2, \qquad (C.48)$$

Appendix D. Fermionic Fock Space

D.1. Introduction

Remember that the harmonic-oscillator-like bosonic commutation relations[69]

$$[\hat{a}_{\alpha},\hat{a}_{\beta}]=0,[\hat{a}^{\dagger}_{\alpha},\hat{a}^{\dagger}_{\beta}]=0,[\hat{a}_{\alpha},\hat{a}^{\dagger}_{\beta}]=\delta_{\alpha,\beta}, \qquad (D.1)$$

give rise to bosonic Fock space in which the oscillator modes correspond to single-particle states $|\xi\rangle$. Here, we shall see how the fermionic anti-commutation relations

$$\{\hat{a}_{\alpha},\hat{a}_{\beta}\}=0,\{\hat{a}^{\dagger}_{\alpha},\hat{a}^{\dagger}_{\beta}\}=0,\{\hat{a}_{\alpha},\hat{a}^{\dagger}_{\beta}\}=\delta_{\alpha,\beta}, \qquad (D.2)$$

give rise to fermionic Fock space. Again, the modes ξ correspond to single-particle quantum states.

Through certain quantum organized transformations (*Bogoliubov transformations* [111] that are included in the class of *organized transformations in nanotechnology* [89]) we can cross from one space to another. This is very useful for some applications where microscopic effects can be obtained for other microscopic effects that were created, affecting a particle in immediate space [112], for example, quantum communication or superconducting material design to enable diverse processes like magnetic levitation and anti-gravity etc.

On the other side, BCS theory describes superconductivity as a microscopic effect caused by the condensation of Cooper pairs into a boson-like state [113] (*boson link-wave of two electrons*). Considering the anti-commutation relations given in (D.2) for one fermionic mode, we can have

$$\{\hat{a},\hat{a}^{\dagger}\}=1, \qquad (D.3)$$

[69] Second quantizing formalism in quantum mechanics.

then we define new operators $\hat{b},\hat{b}^{\dagger}$, such that

$$\hat{b} = s\hat{a} + t\hat{a}^{\dagger}, \qquad b^{\dagger} = s^{*}\hat{a}^{\dagger} + t^{*}\hat{a}, \qquad (D.4)$$

where the latter is the Hermitian conjugate of the first. The Bogoliubov transformation is a canonical transformation of these operators. To find the conditions on the constants s and t such that the transformation remains canonical, the commutator is expanded

$$[\hat{b},\hat{b}^{\dagger}] = [s\hat{a} + t\hat{a}^{\dagger}, s^{*}\hat{a}^{\dagger} + t^{*}\hat{a}] = \dots$$
$$\dots = (|s|^{2} + |t|^{2})[\hat{a},\hat{a}^{\dagger}], \qquad (D.5)$$

To make the transformation canonical, s and t can be parameterized as

$$s = e^{i\theta_1}\cos r, \qquad t = e^{i\theta_2}\,\text{sen}\ r, \qquad (D.6)$$

The point at which it is necessary to perform a Bogoliubov transform becomes obvious. In mean-field approximation, the Hamiltonian of the system can be written in both cases as a sum of bilinear terms with the original creation and destruction operators involving finite $\langle a_{\alpha}^{+}, a_{\beta}^{+}\rangle$-terms, i.e. one must go beyond the usual Harttree-Fock method [114].

The Hilbert space under consideration is equipped with these operators and henceforth describes a higher-dimensional quantum harmonic oscillator (usually an infinite-dimensional one for our Fock space [115]).

The ground state[70] of the corresponding Hamiltonian is annihilated by all the annihilation operators, that is to say

$$\forall i, \qquad a_i|0\rangle = 0, \qquad (D.7)$$

---

70 The ground state of a quantum mechanical system is its lowest-energy state; the energy of the ground state is known as the zero-point energy (*or energy state of the zero-point field*) of the system. An excited state is any state with energy greater than the ground state. The ground state of a quantum field is usually called the vacuum state.

All excited states are obtained as linear combinations of the ground state excited by some creation operators

$$\prod_{k=1}^{n} a_{i_k}^{\dagger} |0\rangle = 0, \qquad (D.8)$$

One may redefine the creation and annihilation operators through a linear redefinition

$$a'_i = \sum_j (s_{ij} a_j + t_{ij} a_j^{\dagger}), \qquad (D.9)$$

where the coefficients s_{ij} and t_{ij} must satisfy certain rules to guarantee that the annihilation operators and creation operators $a_i'^{\dagger}$, defined by the Hermitian conjugate equation, have the same commutators for bosons and anticommutators for fermions. Equation (D.9) is thus the Bogoliubov transformation, which is the organized transformation to define the particle process for a *Fock space* of an arbitrary number of identical fermions, explaining the superconductivity as a Bose-Einstein condensation.

Here, the ground state annihilated by all a'_i is different from the original ground state $|0\rangle$. They can be viewed as Bogoliubov transformations of one another using the operator-state correspondence. They can also be defined as squeezed coherent states. The BCS wave function is an example of a squeezed coherent state of fermions [116]. This is a special fermionic Fock space that is needed to create superconductivity from boson link-waves of fermionic superconducting Fock space.

D.2. Fermionic Algebra and the Multiple Fermionic Modes

We consider here multiple fermionic creation and annihilation operators $\hat{a}_{\alpha}^{\dagger}$ and $\hat{a}_{\alpha}$ that are Hermitian conjugates of each other and satisfy the anti-commutation relations (D.2). For each mode α, we define the occupation number operator

$$\hat{n}_\alpha := \hat{a}^\dagger_\alpha \hat{a}_\alpha, \qquad (D.10)$$

All these operators commute with each another and, moreover, each $\hat{n}_\alpha$, commutes with the creation and annihilation operators for all other modes $\beta \neq \alpha$. Indeed, using the Leibniz rules for commutators and anti-commutators, we have

$$[E,FD] = [E,F]D + F[E,D] = \{E,F\}D - F\{E,D\}, (D.11)$$
$$[EF,D] = E[F,D] + [E,D]F = E\{F,D\} - \{E,D\}F, (D.12)$$
$$\{E,FD\} = [E,F]D + F\{E,D\} = \{E,F\}D - F[E,D], (D.13)$$
$$\{EF,D\} = E[F,D] + \{E,D\}F = E\{F,D\} - [E,D]F, (D.14)$$

From which we obtain

$$[\hat{n}_\alpha, \hat{a}_\beta] = [\hat{a}^\dagger_\alpha \hat{a}_\alpha, \hat{a}_\beta] = \hat{a}^\dagger_\alpha \{\hat{a}_\alpha, \hat{a}_\beta\} - \{\hat{a}^\dagger_\alpha, \hat{a}_\beta\}\hat{a}_\alpha$$
$$= \hat{a}^\dagger_\alpha \times \mathbf{0} - \delta_{\alpha\beta} \times \hat{a}_\alpha = -\delta_{\alpha\beta} \times \hat{a}_\beta \to 0, \forall \beta \neq \alpha, \qquad (D.15)$$

Additionally

$$[\hat{n}_\alpha, \hat{a}^\dagger_\beta] = [\hat{a}^\dagger_\alpha \hat{a}_\alpha, \hat{a}^\dagger_\beta] = \hat{a}^\dagger_\alpha \{\hat{a}_\alpha, \hat{a}^\dagger_\beta\} - \{\hat{a}^\dagger_\alpha, \hat{a}^\dagger_\beta\}\hat{a}_\alpha$$
$$= \hat{a}^\dagger_\alpha \times \delta_{\alpha\beta} - \mathbf{0} \times \hat{a}_\alpha = +\delta_{\alpha\beta} \times \hat{a}_\beta \to 0, \forall \beta \neq \alpha, \qquad (D.16)$$

and

$$[\hat{n}_\alpha, \hat{a}^\dagger_\beta \hat{a}_\gamma] = \hat{a}^\dagger_\beta [\hat{n}_\alpha, \hat{a}_\gamma] + [\hat{n}_\alpha, \hat{a}^\dagger_\beta]\hat{a}_\gamma = -\hat{a}^\dagger_\beta \times \delta_{\alpha\beta} \hat{a}_\gamma +$$
$$\delta_{\alpha\beta} \hat{a}^\dagger_\beta \times \hat{a}_\gamma = (\delta_{\alpha\beta} - \delta_{\alpha\gamma})\hat{a}^\dagger_\beta \hat{a}_\gamma \to 0, \forall \beta \equiv \gamma, \qquad (D.17)$$

also

$$[\hat{n}_\alpha, \hat{n}_\beta] = [\hat{n}_\alpha, \hat{a}^\dagger_\beta \hat{a}_\beta (\text{for some } \beta)] = 0, \qquad (D.18)$$

The fact is that all $\hat{n}_\alpha$, commute with each other, which allows us to diagonalize all of them at once. This gives us the occupation-number basis of states, that is to say, $|\text{set of all } \hat{n}_\alpha\rangle$ for the whole Hilbert space of the theory. Similar to the bosonic case, we may use the $\hat{a}_\alpha^\dagger$ and $\hat{a}_\alpha$ operators to raise or lower any particular $\hat{n}_\alpha$ without changing the other occupation numbers $\hat{n}_\beta$. This means that all the occupation numbers can take any allowed value independently of each other. However, the only allowed values of the fermionic occupation numbers are 0 and 1; thus multiple quanta in the same mode are not allowed.

Note that for a finite set of N modes, the fermionic Hilbert space has a finite dimension 2^N. This fact is for understanding the ground state degenerations of fermionic fields in some non-trivial backgrounds that have zero-energy fermionic modes, that is to say, for N zero modes are independent of their Hermitian conjugates, since the ground level of the whole QFT has 2^N degenerate states.

D.3. Fermionic Fock Space

Now, suppose that there is an infinite but discrete set of fermionic modes α corresponding to some 1-particle quantum states $|\alpha\rangle$ with wave functions $\psi_\alpha(\vartheta)$. In the vector ϑ we include the spin and other non-spatial quantum numbers in $\vartheta = (x, y, z, spin, etc\ldots)$. In this case, the fermionic Hilbert space is

$$\mathcal{F} = \bigotimes_\alpha \mathcal{H}_{\text{mode}\,\alpha}(\text{spanning}\,|n_\alpha = 0\rangle\text{ and }|n_\alpha = 1\rangle), \qquad (D.19)$$

which has infinite dimensions, and we may interpret this as Fock space or an arbitrary number of identical fermions. This is our study space of organized transformations that we require [89].

Indeed, let

$$\hat{\Xi} = \sum_\alpha \hat{n}_\alpha, \qquad (D.20)$$

and count the net number of fermionic quanta in all the modes

$\Xi = 0,1,2,3,\ldots,\infty$. Reorganizing $\mathcal{F}$ into the eigen-blocks of $\hat{\Xi}$, we have

$$\mathcal{F} = \bigoplus_{\Xi=0}^{\infty} \mathcal{H}_{\Xi} = \mathcal{H}_0 \oplus \mathcal{H}_1 \oplus \mathcal{H}_2 \oplus \mathcal{H}_3 \oplus \ldots, \qquad (D.21)$$

Here, the $\mathcal{H}_0$, block spans a unique state with $\Xi = 0$, namely the vacuum state $|\text{vac}\rangle = |\text{all}\, n_\alpha = 0\rangle$. The $\mathcal{H}_1$, block spans states with a single $n_\alpha = 1$, while all other $n_\beta = 0$. Similar to the bosonic case, we may identify states $|n_\alpha = 1; \text{other}\, n = 0\rangle = \hat{a}_\alpha^\dagger |\text{vac}\rangle$, with the single-particle states $|\alpha\rangle$, and hence the $\mathcal{H}_1$, block of $\mathcal{F}$, with the Hilbert space of a single particle. The $\mathcal{H}_2$, block of $\mathcal{F}$, spans the states

$$|n_\alpha = n_\beta = 1; \text{other}\, n = 0\rangle = \hat{a}_\alpha^\dagger \hat{a}_\beta^\dagger |\text{vac}\rangle, \qquad (D.22)$$

with $\alpha \neq \beta$, and only such states. The fermionic Fock space does not allow states $|n_\alpha = 2; \text{other}\ n = 0\rangle$, with doubly occupied modes. Note that in the state equation (22), exchanging $\alpha \leftrightarrow \beta$, results in the same physical state, but with an opposite sign (because $\hat{a}_\alpha^\dagger$, and $\hat{a}_\beta^\dagger$, anti-commute). To be precise, we define the state

$$|\alpha, \beta\rangle := \hat{a}_\beta^\dagger \hat{a}_\alpha^\dagger |\text{vac}\rangle = -|\beta, \alpha\rangle, \qquad (D.23)$$

which is clearly anti-commutative. Likewise, the $\mathcal{H}_3$, block spans the states

$$|\alpha, \beta, \gamma\rangle = \hat{a}_\gamma^\dagger \hat{a}_\beta^\dagger \hat{a}_\alpha^\dagger |\text{vac}\rangle, \qquad (D.24)$$

for *three different modes* α, β, and γ. Then, the $\mathcal{H}_4$, block spans states

$$|\alpha, \beta, \gamma, \delta\rangle = \hat{a}_\delta^\dagger \hat{a}_\gamma^\dagger \hat{a}_\beta^\dagger \hat{a}_\alpha^\dagger |\text{vac}\rangle, \qquad (D.25)$$

for *four different modes* α,β,γ, and $\delta,$ etc. In all cases, the order of the modes $\alpha,\beta,\gamma,\delta,\ldots$ physically does not matter, but affects the overall sign of the state, that is to say,

$$\begin{aligned}&|\text{any permutation } \alpha,\beta,\ldots,\omega\rangle = \\ &|\alpha,\beta,\ldots,\omega\rangle \times (-1)^{\text{parity of permutation}}, \qquad (D.26)\end{aligned}$$

Thus, each space $\mathcal{H}_{\Xi},$ (for $\Xi \geq 2),$ is a Hilbert space of $\Xi,$ identical fermions. Here, we consider their wave functions.

A system of two identical fermions has an anti-symmetric wave function of two arguments $\psi(\vartheta_1,\vartheta_2) = -\psi(\vartheta_2,\vartheta_1)$. A complete basis for such wave functions comprises anti-symmetric tensor products or single-particle-functions

$$\begin{aligned}\psi_{\alpha\beta}(\vartheta_1,\vartheta_2) &= \frac{\psi_\alpha(\vartheta_1)\psi_\beta(\vartheta_2) - \psi_\beta(\vartheta_1)\psi_\alpha(\vartheta_2)}{\sqrt{2}} \\ &= -\psi_{\alpha\beta}(\vartheta_2,\vartheta_1), \qquad (D.27)\end{aligned}$$

Note that such wave functions are not only anti-symmetric in $\vartheta_1 \leftrightarrow \vartheta_2,$ but also separately anti-symmetric in $\alpha \leftrightarrow \beta,$ that is to say, $\psi_{\beta\alpha}(\vartheta_1,\vartheta_2) = -\psi_{\alpha\beta}(\vartheta_1,\vartheta_2),$ and as such we may identify them as wave functions of two-fermion states $|\alpha,\beta\rangle := \hat{a}^\dagger_\beta \hat{a}^\dagger_\alpha |\text{vac}\rangle = -|\beta,\alpha\rangle \in \mathcal{H}_2.$

Likewise, a wave function of Ξ of identical fermions is totally anti-symmetric in its Ξ arguments

$$\begin{aligned}&\psi_{\beta\alpha}(\vartheta_1,\vartheta_2,\ldots,\vartheta_\Xi) \\ &= \psi(\text{any permutation of } \vartheta_1\vartheta_2,\ldots,\vartheta_\Xi)\times \\ &\qquad (-1)^{\text{parity of permutations}}, \qquad (D.28)\end{aligned}$$

A complete basis for such wave functions obtains from totally anti-symmetrized products of Ξ different single-particle wave functions

$$\psi_{\alpha_1\cdots\alpha_\Xi}(\vartheta_1,\ldots,\vartheta_\Xi)=\frac{1}{\sqrt{\Xi!}}\sum_{(\tilde{\alpha}_1,\ldots,\tilde{\alpha}_\Xi)}^{\substack{\text{all permutations}\\ \text{of } (\alpha_1,\ldots,\alpha_\Xi)}}\psi_{\tilde{\alpha}_1}(\vartheta_1)\times\ldots$$

$$\ldots\times\psi_{\tilde{\alpha}_\Xi}(\vartheta_\Xi)\times(-1)^{\text{parity}}$$

$$=\frac{1}{\sqrt{\Xi!}}\det\begin{vmatrix}\psi_{\alpha_1}(\vartheta_1) & \psi_{\alpha_2}(\vartheta_1) & \cdots & \psi_{\tilde{\alpha}_\Xi}(\vartheta_\Xi)\\ \psi_{\alpha_1}(\vartheta_2) & \psi_{\alpha_2}(\vartheta_2) & \cdots & \psi_{\alpha_\Xi}(\vartheta_2)\\ \vdots & \vdots & \ddots & \vdots\\ \psi_{\alpha_1}(\vartheta_\Xi) & \psi_{\alpha_\Xi}(\vartheta_\Xi) & \cdots & \psi_{\alpha_\Xi}(\vartheta_\Xi)\end{vmatrix}, \qquad (D.29)$$

Here, Slater's determinant is not only anti-symmetric in $(\vartheta_1,\vartheta_2,\ldots,\vartheta_\Xi)$, but also anti-symmetric with respect to the single-particle states $(\alpha_1,\alpha_2,\ldots,\alpha_\Xi)$, and so we may identify it as a wave-function of the Ξ-fermion state

$$\left|\alpha_1,\alpha_2,\ldots\alpha_\Xi\right\rangle=\hat{a}^\dagger_{\alpha_\Xi}\cdots\hat{a}^\dagger_{\alpha_2}\hat{a}^\dagger_{\alpha_1}\left|\text{vac}\right\rangle\in\mathcal{N}_\Xi, \qquad (D.30)$$

To complete the picture of the wave-function of the our *fermionic Fock space*, let us spell out the action of the creation operators $\hat{a}^\dagger_a$, and the annihilation operators $\hat{a}_\alpha$. For any Ξ-fermion state $\left|\Xi;\Psi\right\rangle$, with a totally anti-symmetric wave function $\psi(\vartheta_1,\vartheta_2,\ldots,\vartheta_\Xi)$, the state $\left|\Xi+1;\Psi'\right\rangle=\hat{a}^\dagger_\alpha\left|\Xi;\Psi\right\rangle$, has a totally anti-symmetric function of $\Xi+1$ variables. This is to say

$$\Psi'(\vartheta_1,\vartheta_2,\ldots,\vartheta_{\Xi+1})=\frac{1}{\sqrt{\Xi+1}}\sum_{i=1}^{\Xi+1}(-1)^{\Xi+1-i}\psi_\alpha(\vartheta_i)\times$$

$$\times\psi(\vartheta_1,\vartheta_2\ldots,\not{\vartheta}_i,\ldots,\vartheta_{\Xi+1}), \qquad (D.31)$$

while $\left|\Xi-1;\Psi''\right\rangle=\hat{a}_\alpha\left|\Xi;\Psi\right\rangle$, has a totally anti-symmetric function of $\Xi-1$ variables. This brings us to

$$\Psi''(\vartheta_1,\vartheta_2,\ldots,\vartheta_{\Xi-1}) = \sqrt{\Xi}\int d^3\vartheta_{\Xi}\psi^*_{\alpha}(\vartheta_{\Xi})\times$$
$$\times\psi(\vartheta_1,\ldots,\vartheta_{\Xi-1},\vartheta_{\Xi}), \qquad (D.32)$$

The proof of these formulae is left to our readers.

Thanks to the relations (D.31) and (D.32), the Fock-space formulae for the additive one-body operators work similarly to those of the boson case. If in Ξ-fermion Hilbert spaces, we have the total action

$$\hat{\mathfrak{I}}_{\mathrm{Tot}} = \sum_{i=1}^{\Xi}\hat{\mathfrak{I}}_1(i^{\mathrm{th}}), \qquad (D.33)$$

where each $\hat{\mathfrak{I}}_1(i^{\mathrm{th}})$ acts only on the ith particle, then in the Fock space we have

$$\hat{\mathfrak{I}}_{\mathrm{Tot}} = \sum_{\alpha,\beta}\langle\alpha|\hat{\mathfrak{I}}_1|\beta\rangle\times\hat{a}^{\dagger}_{\alpha}\hat{a}_{\beta}, \qquad (D.34)$$

For example, to free the non-relativistic electrons in a box with $\alpha = (\mathbf{p}, s)$, we have

$$\hat{H}_{\mathrm{Tot}} = \sum_{\mathbf{p},s}\frac{\mathbf{p}^2}{2m}\times\hat{a}^{\dagger}_{\mathbf{p},s}\hat{a}_{\mathbf{p},s}, \qquad (D.35)$$

$$\hat{\mathbf{P}}_{\mathrm{Tot}} = \sum_{\mathbf{p},s}\mathbf{p}\times\hat{a}^{\dagger}_{\mathbf{p},s}\hat{a}_{\mathbf{p},s}, \qquad (D.36)$$

$$\hat{\mathbf{S}}_{\mathrm{Tot}} = \sum_{\mathbf{p},s,s'}\left\langle\frac{1}{2}s'\middle|\mathbf{S}\middle|\frac{1}{2}s\right\rangle\times\hat{a}^{\dagger}_{\mathbf{p},s}\hat{a}_{\mathbf{p},s}, \qquad (D.37)$$

where $\hat{\mathbf{P}}$ and $\hat{\mathbf{S}}$ are the momentum and spin matrices of the electrons. Likewise, the two-body additive operators that act in Ξ-fermion spaces as

$$\hat{B}_{\mathrm{Tot}} = \frac{1}{2}\sum_{i\neq j}\hat{B}_2(i^{\mathrm{th}}, j^{\mathrm{th}}), \qquad (D.38)$$

and in the Fock space, it becomes

$$\hat{B}_{\text{Tot}} = \frac{1}{2}\sum_{\alpha,\beta,\gamma,\delta}\hat{B}_{\alpha,\beta,\gamma,\delta} \times \hat{a}^{\dagger}_{\alpha}\hat{a}_{\beta}a_{\delta}a_{\gamma}, \qquad (D.39)$$

for which it is very important note the order of sub-indices in $\hat{a}^{\dagger}_{\alpha}\hat{a}_{\beta}a_{\delta}a_{\gamma}$.

Furthermore, $\hat{B}_{\alpha,\beta,\gamma,\delta} = (\langle\alpha| \otimes \langle\beta|)B_2(\langle\gamma| \otimes \langle\delta|)$.

For example, spin-blind potential $V_2(\vartheta_1 - \vartheta_2)$, becomes

$$\begin{aligned}\hat{V}_{\text{Tot}} &= \frac{1}{2}\sum_{i\neq j}V_s(\vartheta_i - \vartheta_j) \\ &= \frac{1}{2L^2}\sum_{\mathbf{q}}W(\mathbf{q})\sum_{\mathbf{p}_1,\mathbf{p}_2}\sum_{s_1,s_2}\hat{a}^{\dagger}_{\mathbf{p}_1+\mathbf{q},s_1}\hat{a}_{\mathbf{p}_2-\mathbf{q},s_2}a_{\mathbf{p}_2,s_2}a_{\mathbf{p}_1,s_1}, \qquad (D.40)\end{aligned}$$

where $W(\mathbf{q}) = \int d^3\vartheta V_2(\vartheta)e^{-i\mathbf{q}\vartheta}$.

Note that while the formulae for this operator in the bosonic and fermionic Fock spaces have similar forms, the actual operators are quite different because the two Fock spaces have different algebras of creation and annihilation operators and different quantum states (symmetric vs. anti-symmetric). Thus, the correspondence between bosons and fermions is established by the physical effect of similar $V_2(\vartheta_1 - \vartheta_2)$ potentials in this example. However, this property can be extended to many-body systems using number representation and the second quantization established inductively by (D.28)-(D.32) with the applications that we require.

TECHNICAL NOTATION

$\kappa(\omega_1, \omega_2)$ – Spectral curvature or curvature energy. This is given in V/m^3 [8].

$\mathbb{R}^3$ – Real ordinary space or 3-dimensional real space.

$\mathcal{L}^{H}_{\text{Hall}}$ – Topological vector space representing the Hall-type sensor space (or sensing space).

$\tau(\omega_1, \omega_2)$ – Torsion energy or spectra of torsion. Also given in V/m^3 [4].

$\boldsymbol{u}$ – Unitary vector used in the definition of normal curvature. This is $\boldsymbol{u} = \cos\theta\boldsymbol{e}_1 + \sin\theta\boldsymbol{e}_2$, where $\boldsymbol{e}_1$, and $\boldsymbol{e}_2$, are the principal vectors of the surface in the space.

$S^{\gamma}_{\alpha\beta}$ – Spinor subject to the law of transformation of a coordinate system α to a coordinate system β.

$\mathfrak{F}$ – Fourier transform operator.

ω – Angular frequency given for $\omega = \dfrac{2\pi}{T}, \quad T > 0.$

$s(\omega_1, \omega_2)$ – Spectral spinor surface. This in the quantum case models the gravitational waves related to the field torsion.

$\amalg$– Adjunction symbol. This symbol is used in the gluing of cells where there is a particle in immediate phase space.

Ψ_{α_i} – A model algebra of the ith-class α_i. For example, "all models of airplanes of glider-type of the class airbus".

$\mathbb{M}-$Space-time model of 10-dimensional space constituted as a manifold of a complex Riemannian manifold structure.

$\Omega(\Gamma)-$ Space of possible trajectories or paths in space-time that a particle can follow. $\Omega(\Gamma)\subset\mathbb{M}$, on all the elections of possible paths whose statistical weight corresponds to the one determined by the intention of a field.

$\mathfrak{L}^2(O_c(z(t)),\mu)-$ Measures the space of all electronic propagators of all possible paths that can to take the cognitive process of accumulated experiences in the memory of an intelligence. This space has a Hilbert space structure.

$L^2[\phi]-$ State space in functional analysis. This has a Hilbert-space structure. In dynamics and quantum mechanics this space is the Hamiltonian space.

$\mathfrak{I}-$Field action.

$\phi_i(i=1,2,3,\ldots,n)-$ i^{th}-state of energy of a particle or body.

$\omega_{0i}-$ i^{th}-1-differential form in complex geometry.

$M-$ Riemannian orientable compact manifold.

Ω_L- Chern-Simmons form.

$\mathfrak{R}-$ Equivalence relation.

$\omega(\phi(x))-$ Volume form of the energy state $\phi(x)$.

$\mathcal{I}-$ Exploration operator.

$\mathcal{G}-$ Topological group of operator class.

$D_F(x(t)) -$ Electronic propagator of particle x, with trajectory $x(t)$.

$C_{n,m} -$ Configuration field space.

$\int d(\phi(x)) -$ Quantized trajectories.

$\_ \amalg(\rho, \mathbf{v}) -$ Topological vector space of a fluid of density ρ, and velocity $\mathbf{v}$

$\mathscr{M}_H(G, C)$ - Hitchin moduli-space equivalent to space-time in geometrical correspondence and ramification theory.

$P_N -$ Complex projective space of dimension N.

$\Omega_{ij} -$ Differential forms of curvature in a Kähler manifold.

D-modules–Spaces of differential operators that are elements (germs) of an algebraic sheaf. These modules belong to twisted sheaf of differential operators acting on a smooth algebraic variety.

$K_{\mu}^{ab} -$ Contorsion tensor. This is a curvature in the sense of torsion.

CP-violation–In particle physics, *CP*-violation is a violation of CP-symmetry (or charge conjugation parity symmetry): the combination of C-symmetry (charge conjugation symmetry) and P-symmetry (parity symmetry).

GUT-models–Grand Unified Theories. These theories are theories developed to explain all aspects of strong interactions, weak interactions, and supergravity, etc.

P-symmetry–Symmetry due to parity.

C-symmetry–Means the symmetry of physical laws under a charge-conjugation transformation. Electromagnetism, gravity, and the strong interaction all obey C-symmetry, but weak interactions violate C-symmetry. Charge conjugation is a transformation that switches all particles with their corresponding antiparticles and thus changes the sign of all charges: not only the electrical charges, but also the charges relevant to other forces.

vMSM–This is a minimal re-normalizable extension of the Standard Model by right-handed neutrinos. This model explains neutrino oscillations and provides a candidate for dark matter and the mechanism of baryon number generation in the early universe.

$SO(3) -$ Rotation group of range 3. Its elements are the matrices in 3-dimensional real Euclidean space

$A = \begin{pmatrix} \cos\theta & -\sin\theta & 0 \\ \sin\theta & \cos\theta & 0 \\ 0 & 0 & 1 \end{pmatrix}, \forall \theta \in [0,2\pi].$ This group is a Lie group.

$SO(4) -$ Group of rotations about a fixed point in 4-dimensional real Euclidean space. This is also a LiegGroup.

$SU(2) -$ Special unitary group of degree 2, which is the Lie group of 2×2 unitary matrices with determinant 1. $SU(2) \subset U(2) \subset GL(2,\mathbb{C}).$

$\_\|(\sigma, \mathsf{E}) -$ Topological vector space of a fluid of conductivity σ and electrical field E.

$\mathbb{C} -$ Complex number space. This is the set of points *z* such that

$$z = a + ib, \forall a, b \in \mathbb{R}, \text{é}\, i = \sqrt{-1}.$$

C–Algebraic curve. This is an element of a moduli space in algebraic geometry. In field theory, it is re-interpreted as an element of string theory corresponding to an operator in QFT.

$\bar{\psi}\gamma_5\gamma^\mu\psi$ – Term of the spinor axial current.

$\kappa J^\mu_{(A)} J_{(A)\mu}$ – Term that refers to fermion self-interactions induced by quantum torsion.

I_{NY} – Term that is the Nieh-Yan topological invariant density creating the unique Lorentz invariant torsion.

Mathematical Physics, Modern Physics, and Astrophysics Acronyms and Astronomical Nomenclature

SM-Standard Model. This theory is a relativistic theory of particles. The Standard Model of particle physics is the theory describing three of the four known fundamental forces (electromagnetic, weak, and strong interactions, omitting gravity) in the universe, as well as classifying all known elementary particles.

LSZ-Lehmann-Symanzik-Zimmermann (formula).

TQFT-Topological quantum field theory.

QED-Quantum electrodynamics.

QFT-Quantum field theory.

Timelikes-In mathematical physics, this refers to the causal structure of a Lorentzian manifold, describing the causal relationships between points in the manifold.

Holst term-The Holst action is an equivalent formulation of the Palatini action for General Relativity (GR) in terms of vierbeins (4-D space-time frame field) by adding a part of a topological term (Nieh-Yan), which does not alter the classical equations of motion, as long as there is no torsion

$$S-\frac{1}{2}\int ee_I^{\alpha}e_J^{\beta}(F_{\alpha\beta}^{IJ}-\alpha * F_{\alpha\beta}^{IJ})\equiv\frac{1}{2}\int\int ee_I^{\alpha}e_J^{\beta}(F_{\alpha\beta}^{IJ}-\frac{\alpha}{2}\in_{KL}^{IJ}F_{\alpha\beta}^{IJ}),$$

where e_I^{α}, is the tetrad; e is its determinant (the space-time metrics are recovered from the tetrad by the formula $g_{\alpha\beta}=e_I^{\alpha}e_J^{\beta}\eta_{IJ}$, where η_{IJ}, is the

Minkowski metric); and $F_{\alpha\beta}^{IJ}$, is the curvature considered as a function of the connection $A_{\alpha\beta}^{IJ}$, as follows

$$F_{\alpha\beta}^{IJ} = A_{\alpha\beta}^{IJ} = 2\partial_{[\alpha}A_{\beta]}{}^{IJ} + 2A_{[\alpha}{}^{IK}A_{\beta]K}{}^{J},$$

Here, α is a (complex) parameter and we recover the Palatini action when $\alpha = 0$. This only works in 4-dimensional space. So that this can be of free torsion, the covariant derivative defined by the connection $A_{\alpha\beta}^{IJ}$, when acting on the Minkowski metric η_{IJ} must vanish, implying that the connection is anti-symmetric in its internal indices *I*,*J*.

CP-This refers to charge-parity in the context of particles. *CP*-violation is a violation of CP-symmetry (or charge conjugation parity symmetry): the combination of C-symmetry (charge symmetry) and P-symmetry (parity symmetry). CP-symmetry states that the laws of physics should be the same if a particle is interchanged with its antiparticle (C symmetry) while its spatial coordinates are inverted ("mirror" or P-symmetry).

T-symmetry-Or time reversal symmetry is the theoretical symmetry of physical laws under the transformation of time reversal $T : t \mapsto -t$.

GUT-This means Grand Unified Theory, which is a model in particle physics where, at high energies, the three gauge interactions of the Standard Model, comprising the electromagnetic, weak, and strong forces, are merged into a single force.

WMAP-The Wilkinson Microwave Anisotropy Probe (WMAP) is a NASA probe whose mission is to study the Earth's atmosphere and measure the temperature differences observed in the microwave background radiation—a remnant of the Big Bang (see the figure).

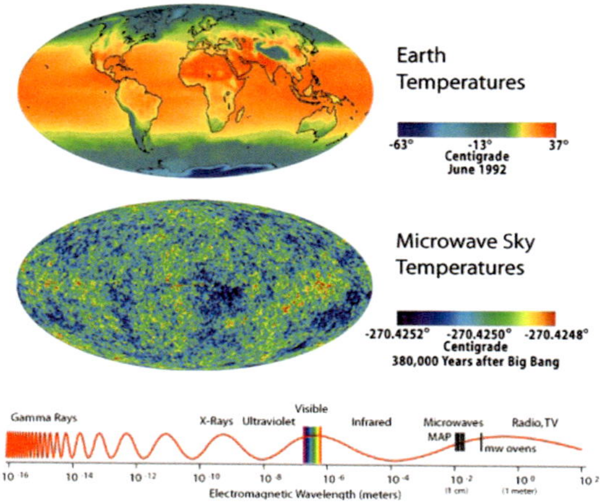

Fig. A) Wmap. B) The wavelengths of radiation detected by WMAP in the microwave region of the electromagnetic spectrum, as depicted in a NASA graph.

Large CMB-Large Cosmic Microwave Background. In Big Bang cosmology, this is electromagnetic radiation that is a remnant of an early stage of the universe, also known as "relic radiation". This is measure covers a large interval of time.

GB Quantization-Gupta-Bleuler quantization, which consists of a QFT formalism, which is a way of quantizing the electromagnetic field.

NASA-National Aero-Space Agency. NASA was founded by German scientists in the United States of America.

SNe-Nearby Type Ia Supernovae. This galaxy-type correlates to Host Galaxy Masses. The SNe-type was discovered by the flux-limited Supernova Legacy Survey (SNLS).

CC-Cosmological Constant.

BAO-Baryon acoustic oscillations. These are fluctuations in the density of visible baryonic matter (normal matter) of the universe, caused by acoustic density waves in the primordial plasma of the early universe. Used in the same way that supernovae provide a "standard candle" for astronomical observations.

CMB- Cosmic Microwave Background.

FLRW-Friedmann-Lemaître-Robertson-Walker metric. We consider the following cases:

i) Flat–FLRW space $(k = 0)$. Considered in Cartesian coordinates and where the Ricci tensor components that are not vanished are

$$R_{tt} = -3\frac{\ddot{a}}{a}, \qquad R_{xx} = R_{yy} = R_{zz} = c^{-2}(a\ddot{a} + 2\dot{a}^2),$$

where the Ricci scalar is

$$R = 6c^{-2}\left(\frac{\ddot{a}(t)}{a(t)} + \frac{\dot{a}^2(t)}{a^2(t)}\right).$$

ii) In more general FLRW space, we consider spherical coordinates, where we have the following unique components of the Ricci tensor

$$R_{tt} = -3\frac{\ddot{a}}{a}, \qquad R_{rr} = \frac{c^{-2}(a(t)\ddot{a}(t) + 2\dot{a}^2(t)) + 2k}{1 - kr^2},$$

$$R_{\theta\theta} = r^2(c^{-2}[a(t)\ddot{a}(t) + 2\dot{a}^2(t)]) + 2k,$$

$$R_{\phi\phi} = r^2(c^{-2}[a(t)\ddot{a}(t) + 2\dot{a}^2(t)]) + 2k)\sin^2\theta,$$

and the Ricci scalar is

$$R = 6\left(\frac{\ddot{a}(t)}{c^2 a(t)} + \frac{\dot{a}^2(t)}{c^2 a^2(t)} + \frac{k}{a^2(t)}\right).$$

vev-Vacuum expectation values.

REFERENCES

[1] Bulnes F, Stropovsvky Y, Rabinovich I. Curvature Energy and Their Spectrum in the Spinor-Twistor Framework: Torsion as Indicium of Gravitational Waves. Journal of Modern Physics. 2017, 8, 1723-1736. DOI:10.4236/jmp.2017.810101

[2] Bulnes F. Gravity, curvature and energy: Gravitational field intentionality to the cohesion and union of the universe. In Gravity Geoscience Applications, Industrial Technology and Quantum Aspect, 1st ed.; Zouaghi T, editor. Intech: London, UK, 2017. DOI: 10.5772/intechopen.71037. Available from: https://www.intechopen.com/books/gravitygeoscience-applications-industrialtechnology-and-quantum-aspect/gravity-curvature-and-energygravitational-field-intentionality-to-thecohesion-and-union-of-the-uni

[3] Dr. Francisco Bulnes, A J. C. García-Limón, L. A. Ortiz-Dumas and V. A. Sánchez-Suarez. Detector of Torsion as Field Ob-servable and Applications. Amer. J. Electrical and Electronic Engineering. 2020; 8 (4):108-115. Doi: 10.12691/ajeee-8-4-2

[4] Penrose, R. Rindler W. Two-Spinor Calculus and Relativistic Fields: Volume 1, 1st ed; Cambridge University Press, United Kingdom, 1984; https://doi.org/10.1017/CBO9780511564048

[5] Bulnes, F. Martínez, I. Zamudio, O. Navarro, E. (2020) Kinematic-Energy Measurements of the Torsion Tensor in Space-Time. In: Advances on Tensor Analysis and their Applications, 1st ed.; Bulnes, F., Ed., Intech, https://www.intechopen.com/online-first/kinematic-energy-measu Bulnes F, Editor. Intech: London, UK, 2020. DOI: 10.5772/intechopen.92815

[6] M. A. Ramírez. I. Ramírez, O. Ramírez, F. Bulnes, Field Ramifications: The Energy-Vacuum Interaction that Produces Movement, Journal on Photonics and Spintronics, 2 (4), 2013, pp4-11.

[7]. Ellis, John R.; Hagelin, J.S.; Nanopoulos, Dimitri V.; Olive, Keith A.; Srednicki, M. (July 1983). \`Supersymmetric Relics from the Big Bang". Nucl. Phys. B238 (2): 453{476.

[8] Jungman G, Kamionkowski M, Griest K. Supersymmetric dark matter. Phys. Rep. 267 5 {(6): 195}373.

[9] F. Bulnes, Baryongenesis until dark matter: H-particles proliferation, MathPhys 9, Belgrade, Serbia, Proc. Phys-Math, September, 2017.

[10] Bulnes, F. (2017) Detection and Measurement of Quantum Gravity by a Curvature Energy Sensor: H-States of Curvature Energy, Recent Studies in Perturbation Theory. In: Uzunov, D., Ed., InTech https://www.intechopen.com/books/recent-studies-in-perturbation-theory/detection-and-measurement-of-quantum-gravity-by-a-curvature-energy-sensor-h-states-ofcurvature-ener https://doi.org/10.5772/68026

[11] Bulnes F, Martínez I, Zamudio O. Fine curvature measurements through curvature energy and their gauging and sensoring in the space. In: Yurish SY, editor. Spain: Advances in Sensors Reviews 4, IFSA; 2016.

[12] F. Bulnes, Research of Curvature on Homogeneous Spaces, TESCHA, State Government of Mexico, 2010.

[13] F. Bulnes, J. C. García-Limón: https://youtu.be/0s5H8w6JfyA

[14] Bulnes F. Electromagnetic Gauges and Maxwell Lagrangians applied to the determination of curvature in the spacetime and their applications. Journal of Electromagnetic Analysis and Applications. 2012; 4(6):252-266. DOI:10.4236/jemaa.2012.46035

[15] Bulnes F. Gravity, curvature and energy: Gravitational field intentionality to the cohesion and union of the universe. In: Zouaghi T, editor. Gravity: Geoscience Applications, Industrial Technology and Quantum Aspect. London, UK: IntechOpen; 20 December 2017. DOI: 10.5772/intechopen.71037. Available from: https://www.intechopen.com/books/gravitygeoscience-applications-industrialtechnology-and-quantum-aspect/gravity-curvature-and-energygravitational-field-intentionality-to-thecohesion-and-union-of-the-uni

[16] Bulnes F, Martínez I, Mendoza A, Landa M. Design and development of an electronic sensor to detect and measure curvature of spaces using curvature energy. Journal of Sensor Technology. Journal of Sensor Technology, 2012;2(3):116-126. DOI: 10.4236/jst.2012.23017

[17] Bulnes F, Martínez I, Zamudio O, Negrete G. Electronic sensor prototype to detect and measure curvature through their curvature energy. Science Journal of Circuits, Systems and Signal Processing. 2015; 4(5):41-54. DOI: 10.11648/j.cssp.20150405.12

[18] F. Bulnes, Transformada de Radon y la Curvatura deun Universo, Tésis de Maestría, UNAM, 2001.

[19] M. Kontsevich, Deformation Quantization of Poisson Manifolds. Lett. Math.Phys. 66 (2003), no. 3, 157-216. 38.

[20] S. Gorbuix, E. Guivenchy, \Torsion Matter and their Corresponding Dirac Equations on Neutrinos and Matter/anti-Matter Asymmetries (Part II: Gravity Insights)," Journal on Photonics and Spintronics, Vol 4, no. 1, 2014, pp15-25.

[21] M. E. Shaposhnikov, G. R. Farrar (1993). \Baryon Asymmetry of the Universe in the Minimal Standard Model".Physical Review Letters 70 (19): 2833-2836.

[22] E. Guivenchy, \Torsion Matter and their Corresponding Dirac Equations on Neutrinos and Matter/anti-Matter Asymmetries (Part I: scattering diference between Neutrinos/anti-Neutrinos)," Journal on Photonics and Spintronics, Vol 3, no. 4, 2014, pp9-18.

[23] É. Cartan (1926), Espaces à connexion affine, projective et conforme, Acta Math. 48: 1–42, doi:10.1007/BF02629755

[24] A-Wollmann Kleinert, F. Bulnes, "Leptons, the subtly Fermions and their Lagrangians for Spinor Fields: Their Integration in the Electromagnetic Strengthening," Journal on Photonics and Spintronics, Vol. 2 no.1 (2013) pp 12-21.

[25] F. Bulnes, "Design of quantum gravity sensor by curvature energy and their encoding," IEEE, Procc. Science and Information Conference 2014, 27-29, Aug., London, UK 10.1109/SAI.2014.6918285.

[26] Wilson, F. L. (1968). "Fermi's Theory of Beta Deacay," American Journal of Physics 36 (12): 1150.

[27] É. Cartan, "Sur une généralisation de la notion de courbure de Riemann et les espaces à torsion." C. R. Acad. Sci. (Paris) 174, 593–595 (1922).

[28] Y. Farzan, S. Pascolli, “A minimal model linking two great mysteries: neutrino mass and dark matter,” The GGI for Theoretical Physics, PRD77 (2008) 43516.

[29] D. R. Peccei, R. H. Quinn (1977). "Constraints imposed by CP conservation in the presence of pseudoparticles".Physical Review D 16 (6): 1791.

[30] M. E. Sevior, A. Ambardar, J. T. Brack, P. Camerini, F. Duncan, J. Ernst, A. Feltham, N. Grion, R. R. Johnson, G. Koch, O. Meirav, D. F. Ottewell, R. Rui, G. R. Smith, V. Sossi, D. Theis, and D. Vetterliz, “Measurement of the near-threshold H(π+,π+π+)n cross section and chiral symmetry,” Phys. Rev. Lett. 66, 2569 (1991)

[31] N. E. Mavromatos, S. Sarkar, “CPT-Violating Leptogenesis induced by Gravitational Defects,” High Energy Physics - Phenomenology (hep-ph); General Relativity and Quantum Cosmology (gr-qc) KCL-PH-TH/2012-23; LCTS/2012-28 arXiv:1211.0968

[32] John Ellis, N.E.M. and Sarben Sarkar. arXiv:1304.5433 [gr-qc]. Phys. Lett. B 725 (2013), 425.

[33] N. E. Mavromatos, S. Sarkar, “Neutrinos in the Early Universe, Kalb-Ramond Torsion and Matter-Antimatter Asymmetry,” High Energy Physics - Phenomenology (hep-ph); High Energy Physics - Theory (hep-th) KCL-PH-TH/2013-45, LCTS/2013-34

[34] Friedman, Alexander (1922), "Über die Krümmung des Raumes", Zeitschrift für Physik A 10 (1): 377–386 arXiv:1312.5230

[35] A. D. Sakharov (1991). "Violation of CP invariance, C asymmetry, and baryon asymmetry of the universe".Soviet Physics Uspekhi 34 (5): 392–393.

[36] M. Shaposhnikov and I. Tkachev, Phys.Lett. B639, 414 (2006), arXiv:hep-ph/0604236 [hep-ph].

[37] M. E. Shaposhnikov, G. R. Farrar (1993). “Baryon Asymmetry of the Universe in the Minimal Standard Model”.Physical Review Letters 70 (19): 2833–2836.

[38] P. Minkowski, Phys, Lett. B 67 (1977) 421.

[39] G. Senjanovic, Phys, Rev.Lett. 44 (1980) 912.

[40] M. Drewes (2013). "The Phenomenology of Right Handed Neutrinos," International Journal of Modern Physics E 22 (8): 1330019. arXiv:1303.6912

[41] E. K. Akhmedov, V. A. Rubakov, A. Y. Smirnov (1998) Phys. Rev. Lett. 81 1359 (Preprint hep-ph/9803255)

[42] U. Debnath, B. Mukhopadhyay, N. Dadhich (2006) Mod.Phys.Lett. A21 399–408 (Preprint hep-ph/0510351)

[43] A. Kostelecky, M. Mewes, "Lorentz and CPT Violation in the Neutrino Sector," Phys. Rev. D70, 031902(R) (2004).

[44] U. Debnath, B. Mukhopadhyay, N. Dadhich, "Space-time curvature coupling of spinors in early universe: Neutrino asymmetry and a possible source of baryogenesis," Mod.Phys.Lett. A21 (2006) 399-408 DOI: 10.1142/S0217732306019542

[45] B. Mukhopadhyay, "Neutrino asymmetry around black holes: Neutrinos interact with gravity," Mod. Phys.Lett. A20 (2005) 2145-2156
DOI: 10.1142/S0217732305017640

[46] Hawking, Stephen W. (1975). "Particle creation by black holes". Communications in Mathematical Physics 43 (3): 199–220.

[47] S. Mercuri, Fermions in Ashtekar-Barbero Connections Formalism for Arbitrary Values of the Immirzi Parameter, Phys. Rev. D 73 (2006) 084016, [gr-qc/0601013]

[48] Klaer, Vincent B.; Moore, Guy D. (2017). "The dark-matter axion mass". Journal of Cosmology and Astroparticle Physics. 2017 (11): 049.

[49] Rubakov, V. A. (2000). "Physical Vacuum: Theory, Experiment, Technology," Physical-Uspekhi 170 (3): 351-352.

[50] F. Bulnes (Ed.), Post-Doctoral Seminar on Spectral Curvature and Spectrum of Curvature, Seminar Mexico-Ukraine-Russia, EAE, Germany-Spain, 2017.

[51] D. W. Sciama, "The physical structure of general relativity,"Rev. Mod. Phys. 36, 463-469 (1964).

[52] Andrea Caputo, Pilar Hernandez, Nuria Rius, Leptogenesis from oscillations and dark matter, *The European Physical Journal C,* volume 79, Article number: 574 (2019).

[53] Fujikura, Kohei; Harigaya, Keisuke; Nakai, Yuichiro; Wang, Ruoquan (2021-03-08). "Electroweak-like baryogenesis with new chiral matter". *Journal of High Energy Physics*. 2021 (7): 224. arXiv:2103.05005

[54] Curtin, David; Jaiswal, Prerit; Meade, Patrick (2012-08-01). "Excluding electroweak baryogenesis in the MSSM". *Journal of High Energy Physics*. 2012 (8): 5. arXiv:1203.2932

[55] Croon, Djuna; Howard, Jessica N.; Ipek, Seyda; Tait, Timothy M. P. (2020-03-31). "QCD Baryogenesis". Physical Review D. 101 (5): 055042. arXiv:1911.01432

[56] Griffiths, David J. (2008). *Introduction to elementary particles* (2nd ed.). Weinheim [Germany]: Wiley-VCH. ISBN 978-3-527-40601-2.

[57] V.M. Abazov; et al. (2010). "Evidence for an anomalous like-sign dimuon charge asymmetry". Physical Review D. 82 (3): 032001. arXiv:1005.2757

[58] Bass, Steven D.; De Roeck, Albert; Kado, Marumi (2021-04-14). "The Higgs boson implications and prospects for future discoveries". Nature Reviews Physics. 3 (9): 608–624. arXiv:2104.06821

[59] Sarkar, Utpal (2007). *Particle and astroparticle physics*. CRC Press. p. 429. ISBN 978-1-58488-931-1.

[60] Frieman, J. A., Particle creation in homogeneous spacetimes. Phys. Rev. D(39), 389 (1989).

[61] J. Schwinger, "Particles, Sources, and Fields," Vol. (1), 3th edition Advanced Book Program, Perseus Books, MA, USA, 1998.

[62] C. Altafini, "Controllability of quantum mechanical systems by root space decomposition of su(N)," Journal of Mathematical Physics 43(5):2051-2062, 2002.

[63] M. Alonso, E. Finn, "Fundamental University Physics, Volume III: Quantum Statistical Physics," Addison- Wesley Publishing Co. MA, USA, 1968

[64] B. T. Holstein, "Topics in Advanced Quantum Mechanics," CA, USA: Addison-Wesley Publishing Company; 1992.

[65] R. P. Feynman, Space-Time Approach to Non-Relativistic Quantum Mechanics, Rev. Mod. Phys. 20, 367-387 (1948).

[66] H. B. Lawson, M. L. Michelsohn, "Spin Geometry, "Princeton University Press; 1989.

[67] A-Wollmann-Kleinert, F. Bulnes, "Leptons, the subtly Fermions and their Lagrangians for Spinor Fields: Their Integration in the Electromagnetic Strengthening," Journal on Photonics and Spintronics Vol. 2, no. 1, Feb. 2013, pp12-21.

[68] A-Wollmann-Kleinert, F. Bulnes, "Spinor Technology: the Field Formalism to the Duality between Quantum Particles and their Waves," Journal on Photonics and Spintronics Vol. 2, no. 2, May 2013, pp9-18.

[69] F. Bulnes, Investigación de Curvatura sobre Espacios Homogéneos, 1era Edición TESCHA, Estado de México, 2010; pp44-66.

[70] F. Bulnes, Detection of finite energy signals to the measurement of the Universe curvature, Iberomet VII, Cancún, Quintana Roo, Mexico, 2002, pp825-843.

[71] F. Bulnes, Integral Theory of the Universe, Proc. Applied Math 2, IMUNAM, ESIME-IPN, Mexico City, 2006, pp73-122.
[72] F Bulnes, *Integral Geometry Methods in the Geometrical Langlands Program*, SCIRP, USA, 2016.

[73] F. Bulnes, "DUAL REPRESENTATION OF THE CURVATURE IN A HILBERT SPACE: CURVATURE AND INTEGRAL TRANSFORMS," JP Journal of Geometry and Topology, Vol. 26, (1), Pages 39 - 51 (January-June 2021) http://dx.doi.org/10.17654/GT026010039

[74] F. Bulnes, "Quantum Gravity Sensor by Curvature Energy: their Encoding and Computational Models", IEEE Proc., SAI2014, London, UK, 2014, pp. 855-861.

[75] Virrey et al, astro-ph/0507170.

[76] WMAP+ Riess et al, 2004 ApJ, 607, 665.

[77] M. Malquarti and A. R. Liddle, Phys. Rev. D66, 023524 (2002), astro-ph/0203232

[78] J. Martin and M. A. Musso, Phys. Rev. D71, 063514 (2005), astro-ph/0410190.

[79] Ph. Brax, J.Martin, "Quintessence and supergravity," *Physics Letters B* Volume 468, Issues 1–2, 25 November 1999, Pages 40-45.

[80] H. Thrift, Y. Stropovsvky, A Mathematician's Search for Technologies of Understanding the Universe, Dr. Francisco Bulnes (Tribute) Cambridge Scholars Publishing, United Kingdom, 2020.

[81] Francisco Bulnes, JC García-Limón, Víctor Sánchez, LA Ortiz-Dumas. Electromagnetic plasma reactor: Implicit application of field torsion I. Int J Electron Microcircuits 2021; 1(1):30-35.

[82] Chen, Francis F. (1984). *Plasma Physics and Controlled Fusion*. Plenum Press. ISBN 978-0-306-41332-2.

[83] Tonks, Lewi (1967). "The birth of "plasma"". *American Journal of Physics*. 35 (9): 857–858.

[84] Morozov, A.I. (2012). *Introduction to Plasma Dynamics*. CRC Press. p. 4–5. ISBN 978-1-4398-8132-3.

[85] Morfill, G. E.; Ivlev, Alexei V. (2009). "Complex plasmas: An interdisciplinary research field". Reviews of Modern Physics. 81 (4): 1353–1404.

[86] F. Bulnes, F. H. Bulnes, E. Hernández, J. Maya. "Diagnosis and Spectral Encoding in Integral Medicine through Electronic Devices Designed and Developed by Path Integrals," Journal of Nanotechnology in Engineering and Medicine ASME, USA, 2011, pp 021009(2) 10.

[87] Feynman, R.P. (2006) [1985]. QED: The Strange Theory of Light and Matter. Princeton University Press. New. Jersey, USA.

[88] Schwinger J (1998) Particles, Sources and Fields. Advanced Book Program, Perseus Books, New York, USA.

[89] Bulnes, F. (2013) Mathematical Nanotechnology: Quantum Field Intentionality. Journal of Applied Mathematics and Physics, 1, 25-44. doi: 10.4236/jamp.2013.15005.

[90] F. Bulnes, "Photonic Chains of Wave-links to the Quantum Communication," Journal on Photonics and Spintronics, (2015) Vol. 4, no. 3, pp10-14.

[91] D. Guderian, Weyl Indices in the Spinor Frame and their Products with Technological Implications," Journal on Photonics and Spintronics, (2014) Vol. 3, no. 1, pp4-11.

[92] Francisco Bulnes (April 3rd 2013). Quantum Intentionality and Determination of Realities in the Space-Time through Path Integrals and Their Integral Transforms, Advances in Quantum Mechanics, Paul Bracken, IntechOpen DOI: 10.5772/53439. Available from: https://www.intechopen.com/books/advances-in-quantum-mechanics/quantum-intentionality-and-determination-of-realities-in-the-space-time-through-path-integrals-and-t

[93] Weylman, H, "Path Integrals for Photons: The Framework for the Electrodynamics Carpet of the Space-Time," *Journal on Photonics and Spintronics*, Vol 1, no. 1, pp21-27.

[94] Faddeev, L. D, and Popov, V. N., "Feynman Diagrams for the Yang-Mills Field", Phys. Lett. B25 (1967) 29.

[95] Jackson, J. D. (2002). "From Lorenz to Coulomb and other explicit gauge transformations". American Journal of Physics 70 (9): 917.

[96] Frampton, P. (2008). Gauge Field Theories (3rd edición). Wiley-VCH.

[97] Gregory S. Adkins, Feynman rules of Coulomb-gauge QED and the electron magnetic moment, Phys. Rev. D36, 1929 (1987).

[98] Gupta, S. (1950), Proc. Phys. Soc. 63A (7): 681–691.

[99] Bleuler, K. (1950), Helv. Phys. Acta (in German) 23 (5): 567–586.

[100] Weyl, H, Raum, Zeit, Materie (Space, Time, Matter), Lectures on General Relativity, in German. Berlin, Springer 1921.

[101] Hall, D. and Wightman, A.S.: A theorem on invariant analytic functions with applications to relativistic quantum field theory, Matematisk-fysiske Meddelelser, 31, 1 (1957).

[102] Wightman, A., Quantum field theory in terms of vacuum expectation values. In: Physical Review. vol. 101, 1956, p. 860

[103] J. Schwinger, "Particles, Sources, and Fields," Vol. (2), 3th edition Advanced Book Program, Perseus Books, MA, USA, 1998.

[104] C. Altafini, "Controllability of quantum mechanical systems by root space decomposition of su(N)," Journal of Mathematical Physics 43(5):2051-2062, 2002.

[105] M. Alonso, E. Finn, "Fundamental University Physics, Volume III: Quantum Statistical Physics," Addison- Wesley Publishing Co. MA, USA, 1968

[106] B. T. Holstein, "Topics in Advanced Quantum Mechanics," CA, USA: Addison-Wesley Publishing Company; 1992.

[107] R. P. Feynman, Space-Time Approach to Non-Relativistic Quantum Mechanics, Rev. Mod. Phys. 20, 367-387 (1948).

[108] H. B. Lawson, M. L. Michelsohn, "Spin Geometry," Princeton University Press; 1989.

[109] A-Wollmann-Kleinert, F. Bulnes, "Leptons, the subtly Fermions and their Lagrangians for Spinor Fields: Their Integration in the Electromagnetic Strengthening," Journal on Photonics and Spintronics Vol. 2, no. 1, Feb. 2013, pp12-21.

[110] A-Wollmann-Kleinert, F. Bulnes, "Spinor Technology: the Field Formalism to the Duality between Quantum Particles and their Waves," Journal on Photonics and Spintronics Vol. 2, no. 2, May 2013, pp9-18.

[111] N. Bogoliubov, On the theory of superfluidity, J. Phys. (USSR), 11, p. 23 (1947).

[112] A. Fetter, J. Walecka: Quantum Theory of Many-Particle Systems, Dover (2003).

[113] L. Cooper (November 1956). "Bound Electron Pairs in a Degenerate Fermi Gas". Physical Review 104 (4): 1189–1190.

[114] J. B. Stochel, (1997). Representation of generalized annihilation and creation operators in Fock space. Universitatis Iagellonicae Acta Mathematica 34: 135–148.

[115] B. Simon, M. Reed. Mathematical methods for physics, Vol. I (functional analysis). New York: Academic Press, 1972.

[116] M. A. Alario, J. L. Vicent, Superconductivity, Eudema Fortuny, Madrid, Spain, 1991.

[117] T. S. choy, 3D Fermi surface Site,
http://www.phys.ufl.edu/~tschoy/r2d2/Fermi/Fermi.html

[118] A. Franklin, Are There Really Neutrinos? An Evidential History (Westview Press, 2004), p. 186

[119] F. Wilczek (2009). "Majorana returns," Nature Physics, 5 (9): 614.

[120] M. Llano, "Unificación de la Condensación de Bose-Einstein con la Teoría BSC de Superconductores," Rev. Ciencias Exactas y Naturais, 2003; vol. 5, no. 1, pp. 9-21.

[121] F. Bulnes, J. Maya and I. Martínez, "Design and Development of Impeller Synergic Systems of Electromagnetic Type to Levitation/Suspension Flight of Symmetrical Bodies," Journal of Electromagnetic Analysis and Applications, Vol. 4 No. 1, 2012, pp. 42-52. Doi: 10.4236/jemaa.2012.41006.

Fig. I. In Slovakia during a mathematics meeting in 2016.

Fig. II. In Frankfurt, Germany, 2019, receiving various honors and awards as doctorate honoris causa in education philosophy and Red Cross badge by ODAEE and members of the Geneva Convention Committee. His merits are his new research method and its implementation at the Research Center in International and Advanced Research in Mathematics and Engineering, founded by him and other distinguished academicians and entrepreneurs.